T0297715

Pattern cutting for clothing using CAD

The Textile Institute and Woodhead Publishing

The Textile Institute is a unique organisation in textiles, clothing and footwear. Incorporated in England by a Royal Charter granted in 1925, the Institute has individual and corporate members in over 90 countries. The aim of the Institute is to facilitate learning, recognise achievement, reward excellence and disseminate information within the global textiles, clothing and footwear industries.

Historically, The Textile Institute has published books of interest to its members and the textile industry. To maintain this policy, the Institute has entered into partnership with Woodhead Publishing Limited to ensure that Institute members and the textile industry continue to have access to high-calibre titles on textile science and technology.

Most Woodhead titles on textiles are now published in collaboration with The Textile Institute. Through this arrangement, the Institute provides an Editorial Board that advises Woodhead on appropriate titles for future publication and suggests possible editors and authors for these books. Each book published under this arrangement carries the Institute's logo.

Woodhead books published in collaboration with The Textile Institute are offered to Textile Institute members at a substantial discount. These books, together with those published by The Textile Institute that are still in print, are offered on the Woodhead website at: www.woodheadpublishing.com. Textile Institute books still in print are also available directly from the Institute's website at: www.textileinstitutebooks.com.

A list of Woodhead books on textiles science and technology, most of which have been published in collaboration with the Textile Institute, can be found towards the end of the contents pages.

Woodhead Publishing Series in Textiles:
Number 137

Pattern cutting for clothing using CAD

How to use Lectra Modaris pattern cutting software

M. Stott

The Textile Institute

Oxford Cambridge Philadelphia New Delhi

Published by Woodhead Publishing Limited in association with The Textile Institute
Woodhead Publishing Limited, 80 High Street, Sawston, Cambridge CB22 3HJ, UK
www.woodheadpublishing.com
www.woodheadpublishingonline.com

Woodhead Publishing, 1518 Walnut Street, Suite 1100, Philadelphia, PA 19102-3406, USA

Woodhead Publishing India Private Limited, G-2, Vardaan House, 7/28 Ansari Road, Daryaganj, New Delhi – 110002, India
www.woodheadpublishingindia.com

First published 2012, Woodhead Publishing Limited

British Library Cataloguing in Publication Data
A catalogue record for this book is available from the British Library.

Library of Congress Control Number: 2012949950

ISBN 978-0-85709-231-1 (print)
ISBN 978-0-85709-709-5 (online)
ISSN 2042-0803 Woodhead Publishing Series in Textiles (print)
ISSN 2042-0811 Woodhead Publishing Series in Textiles (online)

The publisher's policy is to use permanent paper from mills that operate a sustainable forestry policy, and which has been manufactured from pulp which is processed using acid-free and elemental chlorine-free practices. Furthermore, the publisher ensures that the text paper and cover board used have met acceptable environmental accreditation standards.

Typeset by Toppan Best-set Premedia Limited, Hong Kong
Printed by Lightning Source

Contents

Woodhead Publishing Series in Textiles

1 Watson's textile design and colour Seventh edition
Edited by Z. Grosicki
2 Watson's advanced textile design
Edited by Z. Grosicki
3 Weaving Second edition
P. R. Lord and M. H. Mohamed
4 Handbook of textile fibres Vol 1: Natural fibres
J. Gordon Cook
5 Handbook of textile fibres Vol 2: Man-made fibres
J. Gordon Cook
6 Recycling textile and plastic waste
Edited by A. R. Horrocks
7 New fibers Second edition
T. Hongu and G. O. Phillips
8 Atlas of fibre fracture and damage to textiles Second edition
J. W. S. Hearle, B. Lomas and W. D. Cooke
9 Ecotextile '98
Edited by A. R. Horrocks
10 Physical testing of textiles
B. P. Saville
11 Geometric symmetry in patterns and tilings
C. E. Horne
12 Handbook of technical textiles
Edited by A. R. Horrocks and S. C. Anand
13 Textiles in automotive engineering
W. Fung and J. M. Hardcastle
14 Handbook of textile design
J. Wilson
15 High-performance fibres
Edited by J. W. S. Hearle
16 Knitting technology Third edition
D. J. Spencer
17 Medical textiles
Edited by S. C. Anand
18 Regenerated cellulose fibres
Edited by C. Woodings
19 Silk, mohair, cashmere and other luxury fibres
Edited by R. R. Franck

20 Smart fibres, fabrics and clothing
Edited by X. M. Tao
21 Yarn texturing technology
J. W. S. Hearle, L. Hollick and D. K. Wilson
22 Encyclopedia of textile finishing
H-K. Rouette
23 Coated and laminated textiles
W. Fung
24 Fancy yarns
R. H. Gong and R. M. Wright
25 Wool: Science and technology
Edited by W. S. Simpson and G. Crawshaw
26 Dictionary of textile finishing
H-K. Rouette
27 Environmental impact of textiles
K. Slater
28 Handbook of yarn production
P. R. Lord
29 Textile processing with enzymes
Edited by A. Cavaco-Paulo and G. Gübitz
30 The China and Hong Kong denim industry
Y. Li, L. Yao and K. W. Yeung
31 The World Trade Organization and international denim trading
Y. Li, Y. Shen, L. Yao and E. Newton
32 Chemical finishing of textiles
W. D. Schindler and P. J. Hauser
33 Clothing appearance and fit
J. Fan, W. Yu and L. Hunter
34 Handbook of fibre rope technology
H. A. McKenna, J. W. S. Hearle and N. O'Hear
35 Structure and mechanics of woven fabrics
J. Hu
36 Synthetic fibres: nylon, polyester, acrylic, polyolefin
Edited by J. E. McIntyre
37 Woollen and worsted woven fabric design
E. G. Gilligan
38 Analytical electrochemistry in textiles
P. Westbroek, G. Priniotakis and P. Kiekens
39 Bast and other plant fibres
R. R. Franck
40 Chemical testing of textiles
Edited by Q. Fan
41 Design and manufacture of textile composites
Edited by A. C. Long
42 Effect of mechanical and physical properties on fabric hand
Edited by H. M. Behery
43 New millennium fibers
T. Hongu, M. Takigami and G. O. Phillips
44 Textiles for protection
Edited by R. A. Scott

45 Textiles in sport
Edited by R. Shishoo

46 Wearable electronics and photonics
Edited by X. M. Tao

47 Biodegradable and sustainable fibres
Edited by R. S. Blackburn

48 Medical textiles and biomaterials for healthcare
Edited by S. C. Anand, M. Miraftab, S. Rajendran and J. F. Kennedy

49 Total colour management in textiles
Edited by J. Xin

50 Recycling in textiles
Edited by Y. Wang

51 Clothing biosensory engineering
Y. Li and A. S. W. Wong

52 Biomechanical engineering of textiles and clothing
Edited by Y. Li and D. X-Q. Dai

53 Digital printing of textiles
Edited by H. Ujiie

54 Intelligent textiles and clothing
Edited by H. R. Mattila

55 Innovation and technology of women's intimate apparel
W. Yu, J. Fan, S. C. Harlock and S. P. Ng

56 Thermal and moisture transport in fibrous materials
Edited by N. Pan and P. Gibson

57 Geosynthetics in civil engineering
Edited by R. W. Sarsby

58 Handbook of nonwovens
Edited by S. Russell

59 Cotton: Science and technology
Edited by S. Gordon and Y-L. Hsieh

60 Ecotextiles
Edited by M. Miraftab and A. R. Horrocks

61 Composite forming technologies
Edited by A. C. Long

62 Plasma technology for textiles
Edited by R. Shishoo

63 Smart textiles for medicine and healthcare
Edited by L. Van Langenhove

64 Sizing in clothing
Edited by S. Ashdown

65 Shape memory polymers and textiles
J. Hu

66 Environmental aspects of textile dyeing
Edited by R. Christie

67 Nanofibers and nanotechnology in textiles
Edited by P. Brown and K. Stevens

68 Physical properties of textile fibres Fourth edition
W. E. Morton and J. W. S. Hearle

69 Advances in apparel production
Edited by C. Fairhurst

70 Advances in fire retardant materials
Edited by A. R. Horrocks and D. Price

71 Polyesters and polyamides
Edited by B. L. Deopura, R. Alagirusamy, M. Joshi and B. S. Gupta

72 Advances in wool technology
Edited by N. A. G. Johnson and I. Russell

73 Military textiles
Edited by E. Wilusz

74 3D fibrous assemblies: Properties, applications and modelling of three-dimensional textile structures
J. Hu

75 Medical and healthcare textiles
Edited by S. C. Anand, J. F. Kennedy, M. Miraftab and S. Rajendran

76 Fabric testing
Edited by J. Hu

77 Biologically inspired textiles
Edited by A. Abbott and M. Ellison

78 Friction in textile materials
Edited by B. S. Gupta

79 Textile advances in the automotive industry
Edited by R. Shishoo

80 Structure and mechanics of textile fibre assemblies
Edited by P. Schwartz

81 Engineering textiles: Integrating the design and manufacture of textile products
Edited by Y. E. El-Mogahzy

82 Polyolefin fibres: Industrial and medical applications
Edited by S. C. O. Ugbolue

83 Smart clothes and wearable technology
Edited by J. McCann and D. Bryson

84 Identification of textile fibres
Edited by M. Houck

85 Advanced textiles for wound care
Edited by S. Rajendran

86 Fatigue failure of textile fibres
Edited by M. Miraftab

87 Advances in carpet technology
Edited by K. Goswami

88 Handbook of textile fibre structure Volume 1 and Volume 2
Edited by S. J. Eichhorn, J. W. S. Hearle, M. Jaffe and T. Kikutani

89 Advances in knitting technology
Edited by K-F. Au

90 Smart textile coatings and laminates
Edited by W. C. Smith

91 Handbook of tensile properties of textile and technical fibres
Edited by A. R. Bunsell

92 Interior textiles: Design and developments
Edited by T. Rowe

93 Textiles for cold weather apparel
Edited by J. T. Williams

94 Modelling and predicting textile behaviour
Edited by X. Chen
95 Textiles, polymers and composites for buildings
Edited by G. Pohl
96 Engineering apparel fabrics and garments
J. Fan and L. Hunter
97 Surface modification of textiles
Edited by Q. Wei
98 Sustainable textiles
Edited by R. S. Blackburn
99 Advances in yarn spinning technology
Edited by C. A. Lawrence
100 Handbook of medical textiles
Edited by V. T. Bartels
101 Technical textile yarns
Edited by R. Alagirusamy and A. Das
102 Applications of nonwovens in technical textiles
Edited by R. A. Chapman
103 Colour measurement: Principles, advances and industrial applications
Edited by M. L. Gulrajani
104 Fibrous and composite materials for civil engineering applications
Edited by R. Fangueiro
105 New product development in textiles: Innovation and production
Edited by L. Horne
106 Improving comfort in clothing
Edited by G. Song
107 Advances in textile biotechnology
Edited by V. A. Nierstrasz and A. Cavaco-Paulo
108 Textiles for hygiene and infection control
Edited by B. McCarthy
109 Nanofunctional textiles
Edited by Y. Li
110 Joining textiles: Principles and applications
Edited by I. Jones and G. Stylios
111 Soft computing in textile engineering
Edited by A. Majumdar
112 Textile design
Edited by A. Briggs-Goode and K. Townsend
113 Biotextiles as medical implants
Edited by M. King and B. Gupta
114 Textile thermal bioengineering
Edited by Y. Li
115 Woven textile structure
B. K. Behera and P. K. Hari
116 Handbook of textile and industrial dyeing. Volume 1: Principles, processes and types of dyes
Edited by M. Clark
117 Handbook of textile and industrial dyeing. Volume 2: Applications of dyes
Edited by M. Clark

118 Handbook of natural fibres. Volume 1: Types, properties and factors affecting breeding and cultivation
Edited by R. Kozlowski
119 Handbook of natural fibres. Volume 2: Processing and applications
Edited by R. Kozlowski
120 Functional textiles for improved performance, protection and health
Edited by N. Pan and G. Sun
121 Computer technology for textiles and apparel
Edited by J. Hu
122 Advances in military textiles and personal equipment
Edited by E. Sparks
123 Specialist yarn and fabric structures
Edited by R. H. Gong
124 Handbook of sustainable textile production
M. I. Tobler-Rohr
125 Handbook of natural fibres Volume 1: Types, properties and factors affecting breeding and cultivation
Edited by R. Kozłowski
126 Handbook of natural fibres Volume 2: Processing and applications
Edited by R. Kozłowski
127 Industrial cutting of textile materials
I. Viļumsone-Nemes
128 Colour design: Theories and applications
Edited by J. Best
129 False twist textured yarns
C. Atkinson
130 Modelling, simulation and control of the dyeing process
R. Shamey and X. Zhao
131 Process control in textile manufacturing
Edited by A. Majumdar, A. Das, R. Alagirusamy and V. K. Kothari
132 Understanding and improving the durability of textiles
Edited by P. A. Annis
133 Smart textiles for protection
Edited by R. Chapman
134 Functional nanofibers and applications
Edited by Q. Wei
135 The global textile and clothing industry: Technological advances and future challenges
Edited by R. Shishoo
136 Simulation in textile technology: Theory and applications
Edited by D. Veit
137 Pattern cutting for clothing using CAD: How to use Lectra Modaris pattern cutting software
M. Stott

Disclaimer

The content of this book is the sole responsibility of its author. Lectra SA and any companies of the Lectra Group are not responsible in any way for the content of the book or any representations made therein.

Preface

This book provides a practical and easy-to-use guide to how to make the most of Lectra's Modaris pattern cutting software. It is based on my 40 years as a Pattern Cutter, including 10 years using CAD pattern cutting software such as Modaris. This practical experience has, I hope, put me in a particularly strong position to show you how to use Modaris to its full potential in the classroom and in the workplace. As anyone who has used software knows, there are all sorts of shortcuts, functions etc. that you only really learn by trial and error as you use the software in your day-to-day work. I hope readers will benefit from what I have learnt over the years in making my own work easier.

I am going to assume that you already have some computer skills, as it is very unlikely that you have embarked on pattern cutting with CAD without having some knowledge of 'Word' or spreadsheets such as 'Excel', or of using email or the web. Opening a specific program, creating, saving and using files is pretty much the same for all types of computer software. In fact I have discovered that some commands are the same for a number of completely different and unrelated programs. The use of control + Z, for example, seems to be universal in undoing the last command, and the way that functions can have 'hot keys' is not special to Modaris. I will build on this basic understanding but assume you are new to the specific functions in Modaris.

My training courses gave excellent instruction on what the various functions did, but I often forgot how to do something if there was a long time gap between the training and using a function. Because of this I made notes and then refined them every time I needed to perform a specific task, making the task the focus rather than the capabilities of the function. This book began as this series of notes, initially for my own use, but then for Junior Pattern Cutters working on CAD and visiting students on work experience. I hope you find them as useful as I have.

M. Stott

Chapter 1

The role of patterns in clothing manufacture: generating and communicating information

Abstract: A pattern is more than just a template for cutting cloth. It is the document from which a production run of garments is made and used in many different ways by Production Pattern Cutters and Graders, Lay Planners and Factory Manufacturing Personnel. This chapter tells the story of the pattern from design sketch to production and explains the lines of communication, how to generate information, store instructions and pass them on.

Key words: Pattern Cutter, specifications, measurement charts, Lectra Modaris.

1.1 Introduction: the Pattern Cutter's place in apparel production

The thrill of pattern making is the progression of turning a sketch or idea into a set of templates and instructions from which any number of the same garment can be made and, of course, then bought and worn. Embedded into this pattern is an international language of nips and seam allowances, grain lines and balance marks, alluding to the fact that the pattern isn't the end in itself, but just the beginning.

The implications and journey of your pattern, if you think about its life after leaving your hands, is complex and varied: it will be used to mark shapes onto cloth for cutting; it will be graded to make other sizes; it will be referred to in order to ensure that the right trims are used and the correct assembly followed; and it will be cursed if it is in any way lacking in detail, clarity or accuracy. Your pattern can set the standard for the entire production: a daunting thought.

A good Pattern Cutter, like an Architect, while not actually having to do all the different trades him or herself, needs to know how garment making is done, in what order and with what machinery and resources, and have an appreciation of the totality of the job. It's not just the pattern cutting techniques that you need to know, but the raw materials and manufacturing method as well. Pattern cutting is engineering, designing for function and style and resolving problems on the way. It isn't an isolated occupation: you are working as a part of a team in order to produce beautiful, wearable clothing. Each of the separate occupations involved has its own set of skills, but they all rely on the accuracy of the pattern and its associated information produced by the Pattern Cutter to carry out their jobs well.

1.2 The pattern cutting team

1.2.1 The Pattern Cutter

There are many types of Pattern Cutter. In job advertisements you may read Pattern Cutter, or Designer/Pattern Cutter, Creative Pattern Cutter or Production Pattern Cutter and even Pattern Cutter/Grader. These roles do have points of difference referring mainly to the development stage of the pattern.

A Designer/Pattern Cutter or Creative Pattern Cutter will be working at the innovation stage at the beginning of the design and be more concerned with styling and the shape of a pattern, whereas the Production Pattern Cutter will make sure that the mechanics of the pattern work efficiently and that it is absolutely accurate for use in a production run at a factory. The production pattern is then sent to the Grader.

In reality these distinctions become blurred and good practices of accuracy and integrity should be observed at all stages. Recently I have discovered a new role, that of the Pattern Alteration Maker who only makes the pattern amendments. My own view is that the original Pattern Cutter should make the alterations because they understand the process in making the pattern and can therefore make the appropriate amendments. I was once told by my Pattern Room Manager, during the early years of my career, that any work that I left undone had to be picked up by another. In other words, when a pattern left my hands it should be the very best that I could make it. Being accurate from the outset has never been a bad habit and I offer the same advice to you. Whatever you do, do it well.

1.2.2 The Cutter and Lay Planner

Your immediate next user of the pattern will be the Cutter, the one who lays the pattern onto cloth and cuts it out. This may be a Sample Cutter or, in a factory, just one of the many jobs of the Production Cutter. It may also be an automated cutting machine which will use your pattern directly without the use of a card or paper template first. Whichever type it may be, the Cutter will be using the information you have provided on and with your pattern to make sure the pieces are laid on the correct grain of the fabric, that the correct number of each pattern piece is cut and if they need any further treatment such as applying an interlining.

A Cutter will always lay a pattern onto cloth in such a way as to make the most efficient use of the material. This is called lay planning. A Cutter may also be a Lay Planner, but not necessarily. Today lay planning is most usually done on computers using a special program to ensure the maximum efficiency of cloth usage. In Lectra this is called Diamino. It is essential that patterns made using CAD (computer aided design) have the associated text information correctly added at the pattern stage, as computers use the information exactly as given and will not spot human errors. In Modaris this is called the Variant, which we will look at in greater detail later. It is important to understand that all of the information related to a pattern, decided by and typed in by the Pattern Cutter, must be accurate, as this information will directly affect the work of the Lay Planner.

1.2.3 The Pattern Grader

Once a master pattern and base size chart have been established for the sample size, it is then a matter of applying grading increments to obtain the subsequent sizes. How one grades a style may be repetitive, with the same applications over and over again, but the actual garment style must be taken into consideration and the rules applied with the same logic as when making the initial pattern.

A Grader takes the single size pattern, lets say a size 12, and then makes it smaller or larger to create subsequent sizes, e.g. sizes 8 and 10 and sizes 14 and 16 respectively. How much to add and in what places is a skill on its own as not all styles require exactly the same method of grading. Any inaccuracies in the initial pattern will be amplified during grading, once again highlighting the responsibility of the Pattern Cutter to produce precise patterns. A Pattern Cutter may be a Pattern Grader, and vice versa, but not necessarily as they require quite different training and skills.

1.2.4 *The Sample Machinist*

Here is the real reason for the pattern – a garment is made. If the cloth has been cut well and with all of the notches included, the machinist should be able to put it together without further reference other than a sketch. For a prototype garment, called a first sample, the Sample Machinist will identify any improvements which could be made and advise the Pattern Cutter accordingly. It could be that the fabric has reacted differently to expected when cut out and requires a pattern adjustment, or specialist machinery may be needed, which in turn requires a specific seam allowance for the pattern construction to be made in a certain way. A good Sample Machinist is a Pattern Cutter's best friend, testing the pattern and making ready for production. Note though, that the styling and fit of a garment is a separate stage, and not the responsibility of the Machinist.

1.2.5 *The Garment Technologist*

The quality of a garment is in the hands of the Garment Technologist. Once a sample garment is made and approved for manufacture, there is a series of checks before production can begin. It is the Garment Technologist's job to ensure that these have been done. The full role of the Technologist covers more than the range of this book. Briefly it includes everything to do with the technical production of a garment. Examples include making sure that the garment fits well and adheres to the company size chart, that it is graded correctly, that the fabric meets the company performance standards and that aftercare instructions are suitable. Garment Technologists will also approve the trims, making up methods and finishes and ensure that the correct labels are attached in the correct places. This is not an exhaustive list but shows the scope of the responsibilities of the Garment Technologist.

For the Pattern Cutter the most important relationship with the Garment Technologist is that of the fit of the garment. It is the Technologist who will fit the garment in conjunction with the Designer or Buyer and then ask the Pattern Cutter to amend the pattern according to their requirements. To do this she/he will typically issue a measurement chart called a size specification and a set of written comments to explain the changes required. Different companies arrange these communications in different ways but they all essentially lead to an agreed specification which can be checked against production garments to ensure compliance with the original order.

1.2.6 *Quality Assurance and Quality Control personnel*

Quality Assurance (QA) and Quality Control (QC) personnel have subtly different roles to that of the Garment Technologist and, as with the previous roles, have many common tasks. QC usually takes place at certain designated points in a garment's production, to check that all is well and that it meets the specification requirements. QA is the overall term for ensuring that all critical aspects of production are properly monitored.

1.3 Specifications, measurement charts and other documents

1.3.1 *The specification*

The key document accompanying the pattern is the specification (spec), the chart of measurements against which the finished garment will be checked. In theory, a spec comes after the creation of a pattern and its sample by measuring the finished garment. However, the modern process, especially when the Designer and Pattern Cutter are many miles apart, is to issue a specification first with the design sketch stating the finished measurements required.

If you find yourself making a pattern from a sketch with a detailed measurement specification, use it as a support to the sketch, and if amendments are needed, let the Garment Technologist know. The measurements that you have been given are a guide, intended to help not hinder, but may not be workable. Be bold and, if necessary, make recommendations back.

The specification also contains a lot of additional information that may have an impact on your pattern requirements – labels, hangers, interlinings, shoulder pads etc. The size and positioning of a label can determine the size and shape of a facing. For example a large coat label applied to a front facing on the inside of the front edge will need a facing wide enough to accommodate that label. A small label in the back neck position will require a suitably sized and shaped facing, or none at all.

1.3.2 Measurement charts

The on-going debate on sizing and proportions continues. Body shape, like diet and underwear, has changed over the years. Most now grow taller and bigger than in previous years, but not all. Some have become more athletic, and plastic surgery defies all size charts. We no longer wear wasp-waisted corsets everyday or have bustles to hold out the back of our skirts, but we do have under-wired bras, and some women wear no bras at all. Older people begin to shrink in height and their posture changes. It is therefore not easy to make sweeping generalisations about what the average figure measures. The influences on body shape are as complex as they are varied. Keep an open mind and keep on observing and taking note. Chapter 12 discusses these issues in more detail.

Despite the difficulty in establishing standards, there still needs to be a size chart as a reference point. An example is shown in Fig. 12.3. Verify measurements for yourself, to be appropriate to the customer profile you are making for, and take nothing on someone else's say-so. Experience and observation are two vital tools to your trade.

1.3.3 The fit meeting

If you are a Designer/Pattern Cutter or making garments yourself for a single customer, then the decisions you make about fitting adjustments and cloth selection are entirely your own. In commercial manufacturing many other people have a say. In some situations you may work closely with the Designer and attend fittings where you can see the garment on a body and make your own assessments.

However, it may be that your work is for another company, in another location, where a Buyer or Garment Technologist does the fitting and, sadly, you may never get to see your styles on a live model. Instead, someone else will attend the fitting and send you a set of 'comments' telling you how to amend your pattern. If this is you, take it on the chin. Everyone wants a good garment, so there has to a reason for their 'comments'. Read them completely before making any amendments to the pattern to ensure that you understand fully what is trying to be achieved, and then approach that with your expertise, rather than blindly 'doing what you are told'. Revolutionary? Perhaps, but pattern cutting isn't for the faint-hearted, and I like to know that what I am doing is correct in my own judgement, to obtain the client's requirements.

1.3.4 The design sketch

Fashion drawing will always be done on a figure that gives the illusion of being tall and thin. Many have tried to change this and failed. I have given up. If the design drawing is your own for a pattern that you are going to make, then it doesn't really matter. The idea came from you and you can amend and vary it to suit yourself.

Here is what I would like to say to every non-pattern cutting designer. When you are sending your sketch off to a Pattern Cutter with whom you will have little contact, it will help if you make your sketch as 'readable' as possible. It sounds bizarre, but draw what you mean. If this requires two or three sketches, of different angles or layers, then that is what it takes. An Architect does a plan and multiple elevations for each storey of the building, and you must do likewise. If you are the Pattern Cutter receiving a sketch, it pays to study the sketch and fully understand the brief.

1.3.5 Pattern blocks

Chefs collect recipes and write them down to become a vital and personal reference manual. Artists keep sketchbooks and Writers have their notebooks. I keep my blocks. I have what you might describe as basic blocks. In addition, whenever a trend starts to have an underlying structure that I can identify as being a foundation for many other styles, it becomes a block. A suit jacket, a trench coat, pencil skirt, an a-line mini dress, low-cut jeans, strapless dress or waistcoat are all patterns that I have in a block form so that I can easily make the style changes without having to start from scratch every time I make a new pattern. In this way I have patterns already halfway to finished, that I can call on again and again. These have become a part of my tool kit.

The library of blocks that you amass, made by you so that you know and understand every line and shape, is the most valuable asset of any Pattern Cutter. Keep developing and look after your blocks and patterns. With computerisation, for minor re-styling, copying a complete pattern and then making the required changes becomes a very efficient way of making a pattern very quickly.

1.4 Computerised documentation

The boxed text in the right hand column shows an exchange of emails from a supplier disputing specification measurements. This shows both how a specification is a two-way agreement, and also the potential problems in the inappropriate use of computer pattern cutting.

Dear Gwen,

For the back neck we cannot get 34.3 cm, please double confirm for us, technician says it is fifty something. For the chest it should be 46 cm instead of 46.5 cm, for the CF length, it should be 84 cm instead of 86 cm as you have asked us to drop the neck, thanks. One more thing, I would like to seek your opinion. I just employed a paper pattern technician, he uses computer to make paper pattern before, now he takes 3 days to use hands to make a new style paper pattern. How do you think? For our type of dresses, it is better to use hands or computer to make paper pattern? What is your opinion? In the western world, how do people learn to make paper pattern? Use hands or computer? Thanks

Best regards

Psyche

Dear Psyche,

I've amended the spec against the measurements that you have advised. We learn paper and computer, and it is good to be able to turn your hand to both. 3 days does seem a long time to do a paper pattern, but when he's been with you a bit longer he'll get quicker!! How do your other technicians work?

Regards

Gwen

Dear Gwen,

I sacked the new technician as I don't like him, he tried to express how good he was, he had been to Milan he said. I think he is too slow, my other paper pattern technicians can make one paper pattern each day, if that is a difficult one, surely has to take more time. Our paper pattern technicians say they find computer is not good for evening wear styles. Computer is good for basic styles, how do you think? I have a computer for grading in China, just good for grading and calculate yardage consumption? Thanks!

Yours

Psyche

The question asked these days is – hand or computer? I say both. The computer does not make the pattern, you do. However, the computer does do a lot of the time-consuming procedures very quickly. It also stores thousands of patterns that could fill up vast rooms if made from card, and which are quickly retrievable to print off as many copies as you require. Computers also make it possible to email factory patterns that were formerly cut out in card and sent by courier: an expensive and time-consuming process.

Once you have mastered the basics of pattern cutting by computer, it is straightforward and I would never go back to just card. There is no right or wrong way to make a pattern. If it works and gives the desired effect, it is right for the job. However, there are methods and techniques which others have perfected and have been passed down, and I hope to share these with you.

1.5 Tools of the trade

I am rather obsessive about my tools and no one in my working environment would dare to touch them. I like my shears, pencil, set square and all the small tools acquired over a productive working career to be at hand whenever needed. They are not for sharing, however mean that feels at times, but that's how it is.

Even if you are using a computer pattern-making system, it is still necessary to have some hand tools. An essential item is a good pair of paper shears that are not lent or used for any other purposes at all. They are an extension of your hand, and will gain a character and edge that works just for you. I use my shears infrequently these days, compared to the time when they were in my hand for most of the day, but I need them near to me in my tool tray at the ready.

My pencil is the other essential item. When I made patterns on card I used a ballpoint pen (and had inky hands all the time) because it didn't need sharpening and the pen made a slight indentation into the card, which made it easier to cut. Today I have a propelling pencil with lead refills. Actually I have two: one is a point 5 and the other a point 7. They do slightly different things, but your own experience will explain which is which or if, indeed, you require two. You may not.

Then there is my set square. My oldest one is of course only in inches. Then there is my favourite which is in both inches and centimetres. The newest, only in centimetres, is not used very often, and then mostly by visiting students (I do break my rules). You might choose the Pattern Master, designed by Martin Shoben, a variation that incorporates French curves.

There are an assortment of small tools – notchers, bradle, tracing wheel, Stanley knife, short metal rulers – all of which used to be essential for making card patterns but are seldom used now. There is, of course, my tape measure which hangs around my neck all of the time. With some paper I am all set, nearly. The last item is pattern blocks. I didn't start out with pattern blocks. They have evolved with my career but they have become my defining resource. A selection of blocks is essential. Add to that a measurement chart and it's time to begin.

Essential equipment is: shears, pencil and rubber, set square and tape measure, pattern blocks and measurement charts, masking tape – to tape pattern pieces to digitising board – fabric shears, a few needles and pins. Extras for making card patterns include: notchers (or nips), bradle, tracing wheel, Stanley knife, metre rule, short metal ruler and a few coloured felt tip pens.

Chapter 2

Introduction to the keyboard, mouse and screen layout in Lectra Modaris pattern cutting software

Abstract: The keyboard, screen and mouse work together to provide access to the software commands. This chapter looks at the options and shows how becoming dexterous with your hands results in efficient use of the functions. The chapter covers how to select and move pattern pieces around the screen and in relation to each other. The chapter includes an exercise that uses all of the 'zooming and moving things around' options using **J/j**, **7** and **8**, Home, Current sheet, End, the Page up/Page down keys, zoom functions and marry and divorce.

Key words: Lectra, Modaris, pattern cutting, orientation, keyboard, mouse, screen, *x*-axis, *y*-axis, clockwise, zooming and moving.

2.1 Introduction to Modaris

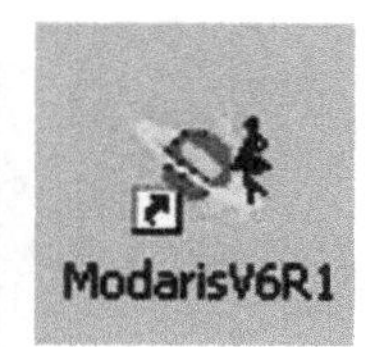

Established in France in 1973, Lectra is a leading company in computer aided design (CAD) software for the textile industry. Launched in 1984, its Modaris software provides on-screen management of pattern design and modification. Modaris is widely used in both the textile industry and by students in colleges of fashion.

2.1.1 Opening Modaris

Modaris has many menus with varying functions that do all sorts of things, all of which in time you will get to know. To begin with you need to orientate yourself with the keyboard and screen. If you haven't already opened Modaris, do this now by double clicking on the program icon.

2.1.2 *Using your hands efficiently*

How you actually use your hands to navigate your way around the different menus will develop in time, but learning some good practice now will make you more efficient. As with most computer systems, there are a number of ways to achieve the same thing. As an example, to display a menu from the right-hand column on the screen, you can select from a box on screen using the mouse or press one of the function keys (**F1** etc.) on the keyboard. Many of the sub-menu functions also have a keyboard equivalent. It makes good sense to learn some of the keyboard actions and employ both hands, so that you don't leave you right (or left) hand to do everything, one finger at a time.

All of your work on the pattern will be done with the mouse, as if you had a pencil in your hand, and it is advisable to keep your right hand (or left if you are left handed) mainly for the mouse. There are times when both hands are needed on the keyboard, in typing for example, but, in general, try keeping the right hand on the mouse. Learn the keyboard commands using your left hand. In this way you will not need to take your hand off the mouse any more than is absolutely necessary, and it will speed up your work.

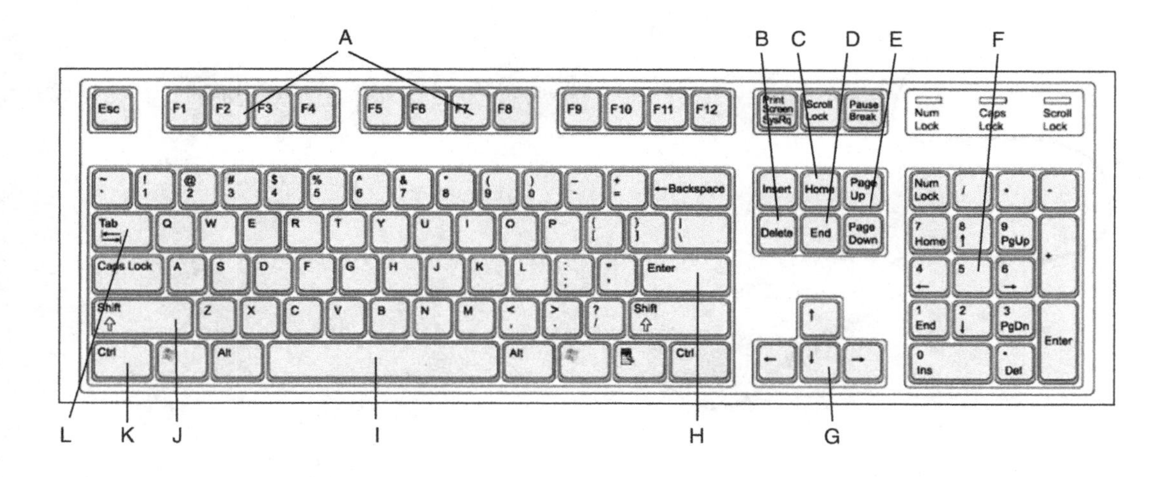

Specific keys

A: **Function** keys (**F1**, **F2** etc.).	Shortcuts to function menus on the screen.
B: **Delete**.	Alternative method of deleting anything. Use with caution.
C: **Home**.	Brings a single selected sheet to full screen.
D: **End**.	Used with the mouse to pick up and move a sheet.
E: **Page up** or **Page down**.	Scroll between successive sheets.
F: **Number keypad**.	Use for entering specific number values.
G: **Movement keys**.	Use for moving between value fields when entering measurements.
H: **Enter**.	Use with mouse to 'click and drag' a rectangle for enlargement.
I: **Space bar**.	The choice bar, when more than one option is available with a function.
J: **Shift**.	Used to vary a function.
K: **Control**.	Used to vary a function.
L: **Tab**.	Used for inserting existing files into the current file. Tab can also be used to display a list of options.

2.2 The keyboard

2.2.1 Typing

The keyboard is, of course, used for actually typing words into the text boxes in the normal manner, but it also has specific commands attached to individual keys. When typing in the normal way, to enter information into text boxes or to complete the **variant**, both hands will be used. When working on a pattern on screen, the most efficient use of your hands will be to keep the left hand to the left side of the keyboard so that there is easy access to the function keys across the top and to **Shift** and **Control** which will be just under your little finger. **J/j** then will be near to your index finger.

It is surprising how often these same keys are used and you will be able to select them instinctively and without looking in a very short time. Most pattern making is done using the colour-coded functions placed in a vertical column on the right of the screen. Select the menu using an F key with your left hand and almost simultaneously the function tool with your mouse.

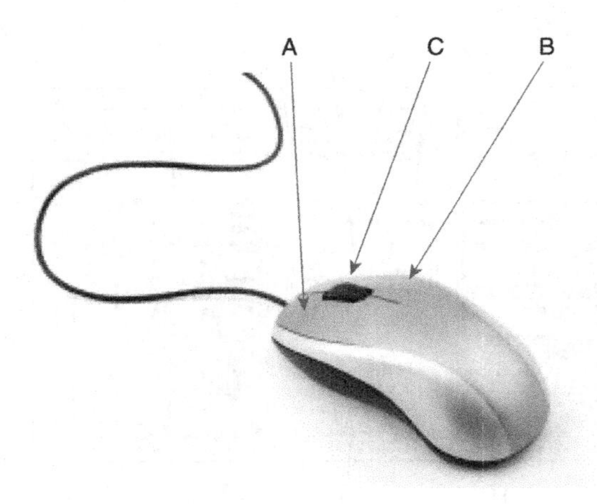

2.3 The mouse

2.3.1 Using the mouse

The mouse, which controls the movement and placing of the cursor, represents your hand moving around the screen to select or perform tasks, and will also become quite instinctive in a very short period of time. Do make sure that your mouse fits your hand and that you have a good, ergonomic mouse mat. You may be holding it for several hours a day. There is a left and right button, and maybe a wheel in the centre. These options are used to begin and end or vary a function.

2.3.2 Mouse buttons

A **Left button** under index finger is used to select an option. Place cursor over the required point and click to select. A single click will open a function, or complete a function. Click and drag is used to select an area or multiple points. Click and hold the button then drag across the area to be selected. Release the button and the selection is highlighted.

B **Right button** under middle finger is generally used to complete a function. Also used to make a selection.

C **Wheel**. When creating a Bezier or semicircular line, the wheel is used to delete the last point created before finishing the line. The wheel is also used to undo the last click when extracting a piece with the seam and cut tool.

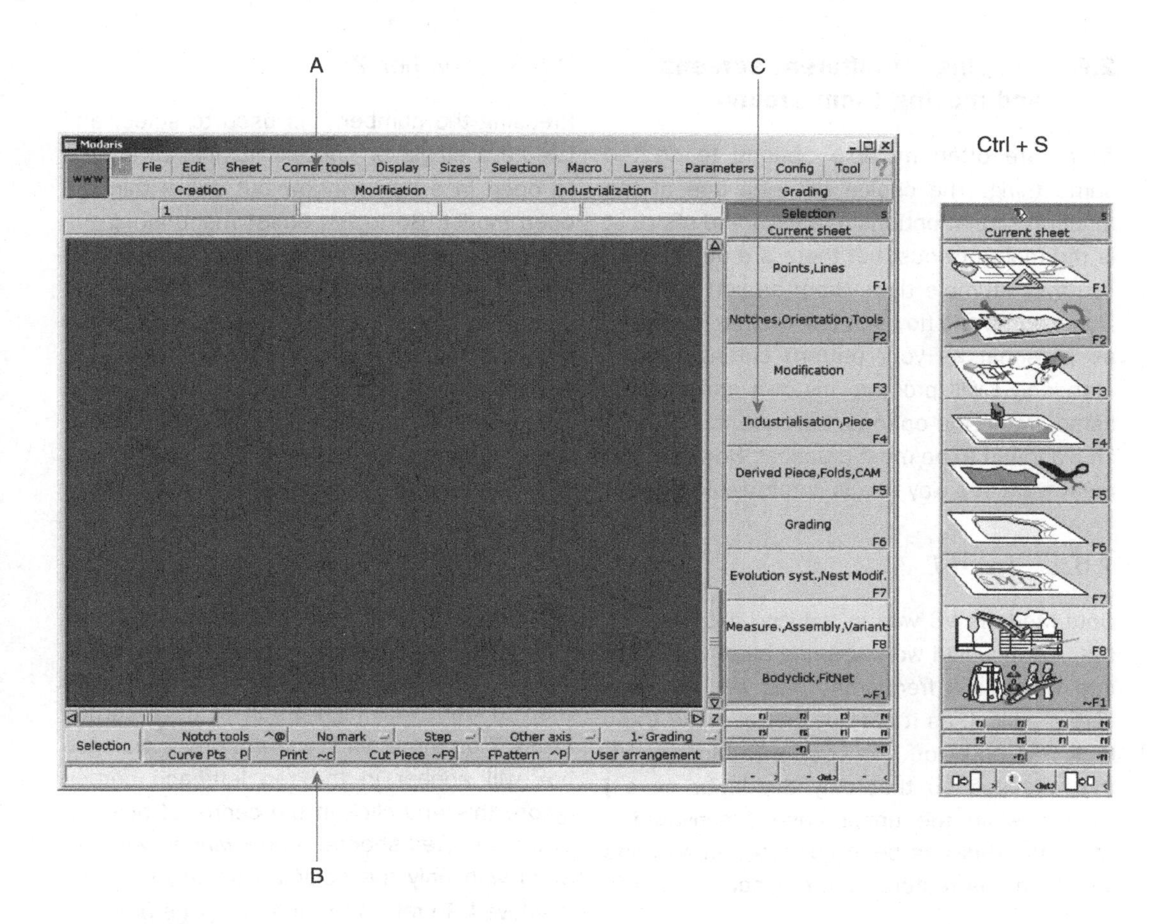

2.4 The Modaris home screen layout

The Modaris home screen has three very distinct areas of choice around the central main working area. Some of these functions have a keyboard command and so do not have to be accessed via the screen. Learning these menus and commands constitutes the main purpose of this book and will be looked at in detail later. Here we look at the main layout and menu headings.

A Along the top are menu headings with the first two familiar to all programs – file and edit. The rest are more specific to Modaris. They will all have drop-down menus for further choice of use.

B Along the bottom of the screen are a series of on/off and simple choice buttons.

C Down the right-hand side are the colour-coded function menu headings. These are the toolbox specific to Modaris. You can choose to display with icons or words by pressing **Ctrl** + **S**.

2.5 Navigating the home screen: the *x*-axis and *y*-axis

The computer screen and digitising board are navigated by use of an '*x*' and '*y*' axis. Many functions operate in a clockwise direction so it makes good sense to always work in a clockwise direction yourself when appropriate. In order to know which way you are going, and in what direction, it is helpful to place a small diagram on both the digitising board and your computer screen until it has become second nature.

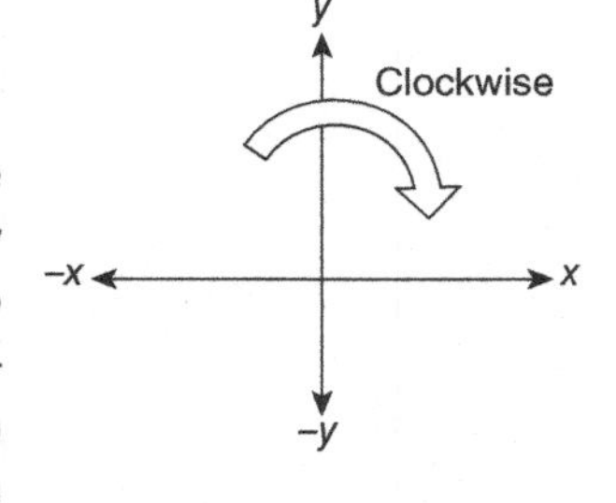

2.6 Bringing up different screens and moving them around

There are often multiple options to do the same thing. The choice between one of the function screen options or a keyboard shortcut is the most obvious, but there are other ways. You may stumble upon them as you work, as I have done, and how you use them will become as personal as your pattern cutting. Understandably I will promote my own preferences, listed first in the options, because this is what I have found to be most practical. However, you must work in a way that is comfortable for you.

2.6.1 *J and j*

Until Modaris V6 was introduced in 2010, **J/j** (i.e. **J** or **j**) would work to bring all of your work into view. As a frequently used key, this was very practical, as it did not matter if the **Caps lock** was on or off. However, with Modaris V6 this changed so that only the lower case **j** worked while the upper case **J** activated a shortcut. This has been corrected in Modaris V7 which has restored the original use of **J/j**.

A major reason for the **J/j** function to remain free of shortcuts is the constant use of these keys, but more specifically when using the **7** to isolate selected sheets. Using the **8** takes you back to all of the sheets, while **J/j** takes you back to just the selected sheets.

2.6.2 *Number 8*

Pressing the number **8** will also bring all work sheets into view. Number **8** is the companion of number **7** which is used to isolate selected sheets – very useful when there are a lot of sheets on screen, and you want to work with just a few of them. **8** takes you back to the full model. **J/j** takes you back to the selected sheets only.

2.6.3 *Number 7*

Pressing the number **7** is used to select and isolate a single or specific number of sheets to open in a new window but still within the open model. Go to the **Sheet** menu along the top of the screen. Choose **Sheet select** from the drop-down menu which will change the cursor icon to a **pointing finger**. Click on the centre of the piece that you want to work with and it will then display small white squares at each corner. To unselect while on **Sheet select**, you can simply left click on the piece. A right click will work as the 'selection' tool and will deselect all the pieces.

Now move your cursor away from the black line that marks the edge of the working page and **right click**. The finger icon will turn to an arrow. You will now have all of the sheets that you wish to work with highlighted at each corner. Press **7** on the main keyboard, and a dialogue box will appear in the top left-hand corner. Ignore this and click in the centre of any one of the selected sheets. A new window will now open with only the selected sheet or sheets displayed. To return to the main page press **8**.

TIP – I often forget a piece and have to go back (press **8**) and then re-select. You will find that all the pieces that you have selected are now in order starting from the top left of the screen, and end at the title page, so re-selecting is relatively straightforward. Married pieces can be selected too.

2.6.4 *Home*

To bring a single sheet to fill the screen, first click on the sheet. It will turn grey to show it is selected. Then press **Home**. It is worth noting that **Current sheet** placed at the top of the tools menus does the same thing.

2.6.5 *End*

Select **End** then left click and release to select and move an individual sheet around on the screen. Click again to place it where required.

2.6.6 *Page up, Page down*

Page up/Page down are used to move between sheets. When checking all of the pieces in order, it is a quick way to scroll through all of the sheets.

2.6.7 *Zoom*

Press **Enter** + **click and drag** with the mouse to form a rectangle on your selected area, as big or small as required. At the base of the tools function menus (right-hand side of screen) are three zoom function keys giving varying zoom sizes. These provide another option for zooming in or out.

2.6.8 *Marry*

To place one pattern piece onto another sheet, select **F8 – Marry**. Decide which sheet you wish to be the host sheet. Then click on a point on the guest sheet first, and then on the corresponding point on the host sheet. The first sheet will be placed into the host sheet and turn red. Note that more than one piece can be married at a time. Useful for checking hem runs, armhole shapes, sleeve heads into armholes etc.

Now use **F8 – Move marriages** to move the guest sheet or sheets around the host sheet.

F8 – Pivot will pivot the piece. Note the piece will pivot about the last selected point, so choose your last **Move marriage** with care.

F8 – Walking pieces allows you to move one piece along the edge of another, as one would 'walk' pieces together with paper patterns. The direction of the 'walking' piece is determined by the way you move the mouse. The **Space bar** is used to change the guest sheet selected.

Shift + **F8 – Marry.** When marrying pieces the addition of **Shift** will turn the piece over.

2.6.9 *Divorce*

F8 – Divorce will separate the pieces. The host sheet will probably now be much bigger than it needs to be, as it remains the size needed to accommodate the married pieces. The command you need to resize a page back to its smallest size is **a** (ensure that **Caps Lock** is not switched on). Normally the function only needs to be used once.

2.7 Practice exercise

Open a model and use the above functions to move around the screen.

Use **Home** to select a single sheet and then **J/j** or **8** to bring everything back into view.

Use the **Enter** + **click and drag** to zoom into small areas and then **J/j** or **8** or **Home** to zoom out again, depending on the choice of a single sheet or the whole model.

Use the **Page up** and **Page down** keys to scroll between sheets.

Use the **End** key to rearrange pieces to suit your preference.

Try selecting some sheets to a new window using **7**, then further zooming in on that window. Use the **J/j** and **8** functions to become familiar with their difference.

Use the **Marry** and **Divorce** functions to manoeuvre pieces together on the same sheet.

Practise all of these actions a few times to get your hands used to the commands which move around the pattern pieces and screens. Try to keep your right hand on your mouse and select the keys with your left. Once you can move things around, actual pattern making will be more straightforward.

Chapter 3

Title blocks in Lectra Modaris pattern cutting software

Abstract: Title blocks are fields that hold text information such as pattern numbers or piece names which will be important for later users of a pattern. Title blocks include the **model identification sheet** as well as individual sheets and the variant which records cutting information. This chapter discusses how to use title blocks effectively.

Key words: title blocks, **model identification sheet**, variant, Lectra Modaris.

3.1 Introduction: information for patterns

As discussed in Chapter 1, the making and using of patterns to produce garments is undertaken by a series of people. The information written on or attached to a pattern becomes vital in the communication from one task to the next. On a card pattern, apart from nips or notches which are cut directly into the card itself, it is a straightforward matter of writing directly onto the pattern all of the information required. It is then read by the next user, who, importantly, is usually able to spot any errors and amend as necessary. Basic information required on a pattern is:

- pattern number,
- size,
- piece name,
- grain line and direction,
- how many to cut,
- fabric type,
- any special information such as fusing or pleating, gathering etc.

On a computer-made pattern this information is embedded into the pattern and extracted by the relevant software when used in another application. The computer program will extract the information exactly as you have entered it and mistakes can easily get overlooked. It is very easy for small errors to go unnoticed and cause no harm until a very much later stage. Errors are usually discovered because of a problem in the manufacturing process, which either cannot be rectified or causes much extra work. Attention to detail then, in every aspect, is an essential part of all pattern making.

3.2 Title blocks

There are four main types of sheets in Lectra Modaris: the model identification sheet; the individual sheet; the variant sheet; and the variant spreadsheet. These sheets are shown in Figs 3.1–3.5. The text boxes on one sheet are linked to other sheets, as well as to other functions throughout the whole of Lectra capabilities. Once you understand these links, the importance of deciding how and why to utilize them becomes significant.

When making patterns on a computer, the variety of information required is held in different ways, but one of the main places is in what are called **Title blocks**. Title blocks are the fields at the left-hand side and along the bottom of a sheet that hold text information.

When plotting a pattern, text will also be printed simultaneously, depending on what is input by you. Some of this will be plotted by default, that is done automatically, and some because of your setting preferences. For instance:

- pattern or model number and size is by default,
- pattern comments are selected according to your preferences,
- piece names will be as you have typed in on individual sheets.

When you are inserting pattern sheets from one model to another (i.e. copy and paste), you will need to look up a list and be able to identify the pieces that you require. If you use well thought-out and consistent labelling, this will be quick and easy. A suggested list is shown later.

When printing out a **variant** sheet, which is a text document and sent to a regular A4 printer, all the information you have typed into the boxes is printed out, probably for another person to use, and so must be accurate.

Marker making in Diamino will extract information from your pattern and display it in the marker-making screen. You may not need this information, but another user will, so it is essential that the information is correct and follows a consistent logic.

Some fields extract their information from the **model identification sheet** and others use information specific to themselves. But don't let this layering baffle you. In time and with use it will all make sense.

3.1 Model identification sheet

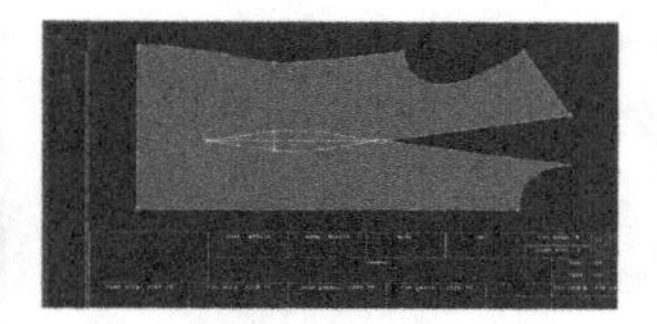
3.2 Individual sheet

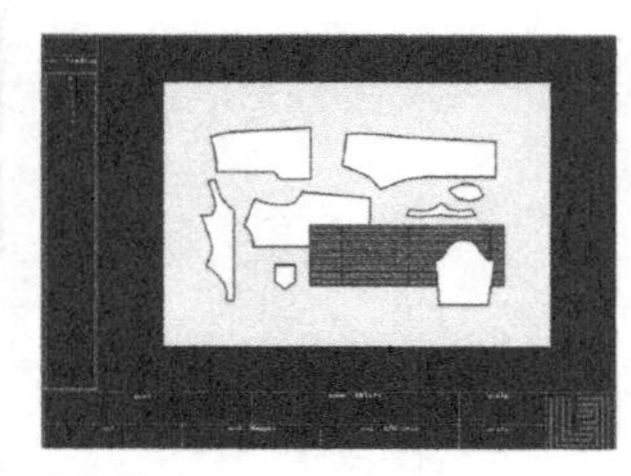
3.3 Variant sheet

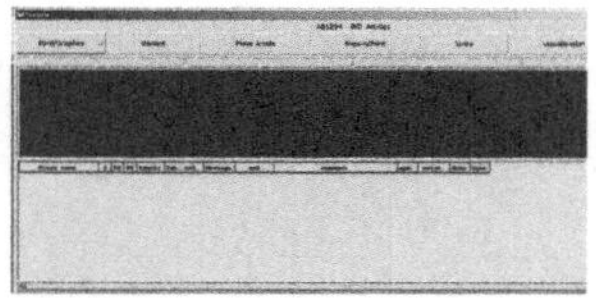
3.4 Variant spreadsheet empty

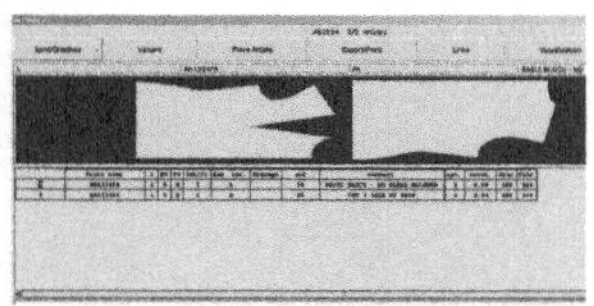
3.5 Variant spreadsheet added

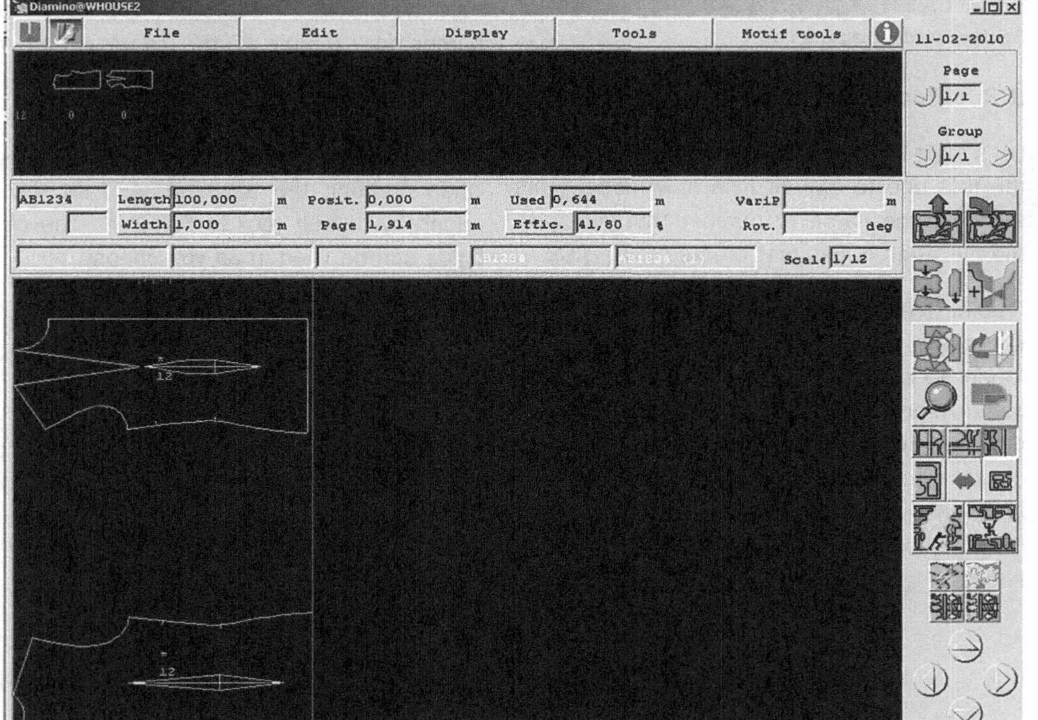

3.6 Diamino sheet

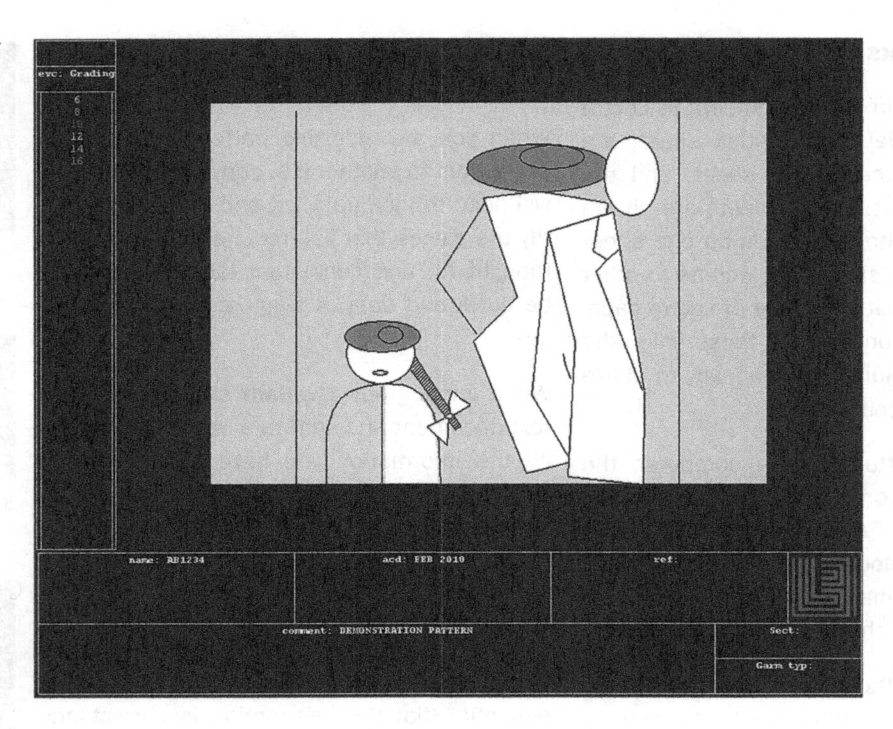

3.7 Model identification sheet

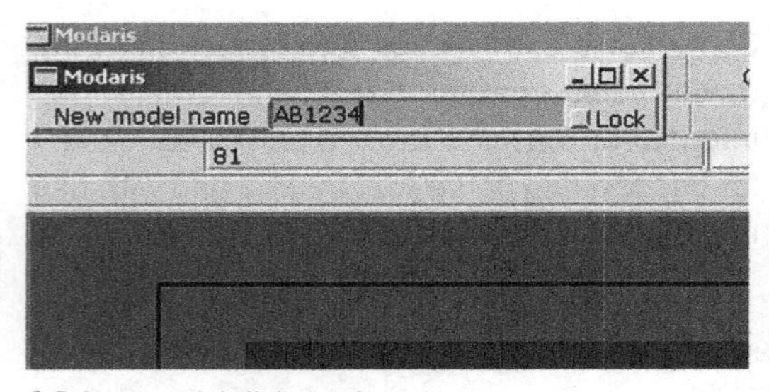

3.8 New model dialogue box

3.3 Model identification sheet

Let's start with the **model identification sheet**, the first one to appear in a new model. This will have the name or number that you gave the file when it was first created (Fig. 3.8).

A point to note is that the **model identification sheet** is a sheet in its own right and does not 'open' to a further level. This sheet contains basic information such as the file name or number and any other information that you decide you need to know about the model. Additionally, it is advisable that the size range is attached before creating further model sheets (Fig. 3.7).

The size range is inserted by use of **F7**, **Imp Ev**, which stands for Import Evolution (left border of Fig. 3.7) When you create a new model, it is important to add the size range at the same time. It took me a long time and much frustration to find out that my patterns would not print out until there was an associated size range, indicated by the list of size names, for example 10–12–14 etc. in the text box on the left-hand side of the sheet.

Name: is the number you gave your new model when creating the new model. This will be added by default (box in Fig. 3.7). The other fields can be filled in as you choose.

Acd: analytical code can have a variety of uses, i.e. the date of creation of model or a collection identifier.

Comment: a description of the model is the most useful information to include here.

Ref: at this stage it is not necessary to fill in this field.

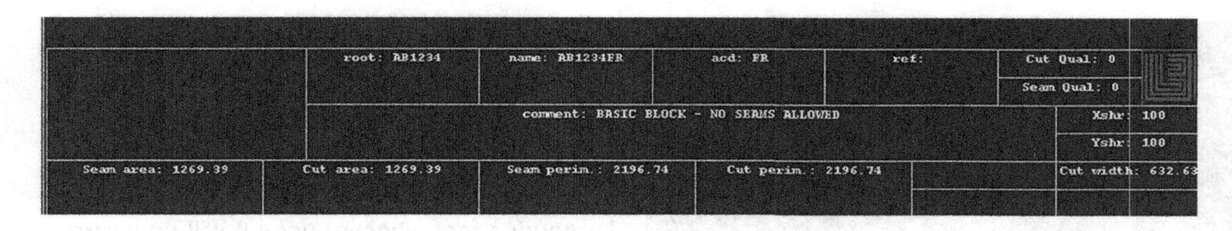

3.9 Individual sheet title blocks

3.4 Individual sheet

Next we come to the **Individual sheet** title blocks. Here we have two identifiers:

- **root:**
- **name:**

Each has different uses (see Fig. 3.9). The **name** of the file is, by default, the same as the number or name that you gave the file when creating the new file. As you create new sheets, this number or name will be displayed automatically as **root** and **name**. You can change the **name** if you wish but not the **root**.

Root: will always refer to the number of the model created when creating a brand new file, and will take its number from the model identification sheet. This is useful because, if you import a piece from another model, it will be identifiable because of this **root**.

Name: this is the same as the pattern within which it was created. If you import a piece, it will retain this number until you change it with the **edit** feature. This is useful because it will identify what has been imported. When a new individual sheet is first created its **name** will be the same as the **root** but with a suffix indicating which number sheet it is. For example, AB1234, the model number will have 1,2,3 etc. added to each new sheet: AB12341, AB12342, AB12343 etc. This can be replaced with a more useful code, FR or BK etc. to become AB1234FR, AB1234BK etc. See the pattern piece names abbreviation list in Appendix 2.

Acd: it is usual to use this analytical code to repeat the code added to the name in the text box. This is useful for those working in **Diamino** to identify pattern pieces. While it will make little difference to you, it will make a difference to others using your work after you.

Comment: this text box has many potential uses, and best use will depend upon company policy and use of patterns. What is written in this box can be plotted out onto patterns, and incorrect information can lead to errors and confusion. Let's look at the options.

It is possible to plot out patterns with very little information, perhaps just the model number and size. If the pattern is for your own use this may be sufficient. However, my experience is that, when I come back to a pattern after some time, I have forgotten the details. In addition, if I pass the pattern onto someone else, it is essential that they can use the pattern without referring back to me. As a result, the extra information added to a pattern becomes a vital part of its integrity.

Grain lines and notches are a part of that information and are added separately, but how many to cut and of what, and whether a single or pair become essential details. This is where the comment box is useful.

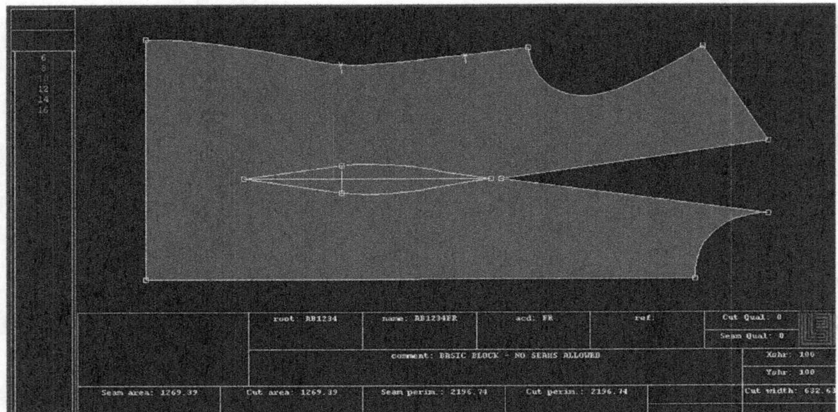

3.10 *Basic block without any seam allowance*

Example 1 (see Fig. 3.10 and Fig. 3.11) – The comment box tells me that this is a basic block and that no seam allowances are added. When this pattern is plotted out, this text will be printed onto the pattern, ensuring that whoever uses it will be aware that it is a net block pattern.

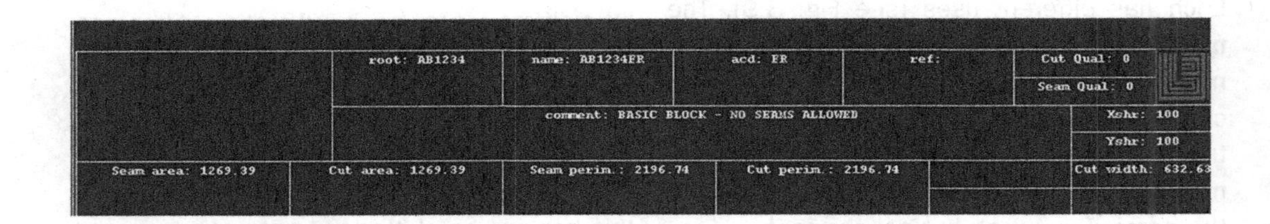

3.11 *Comments box*

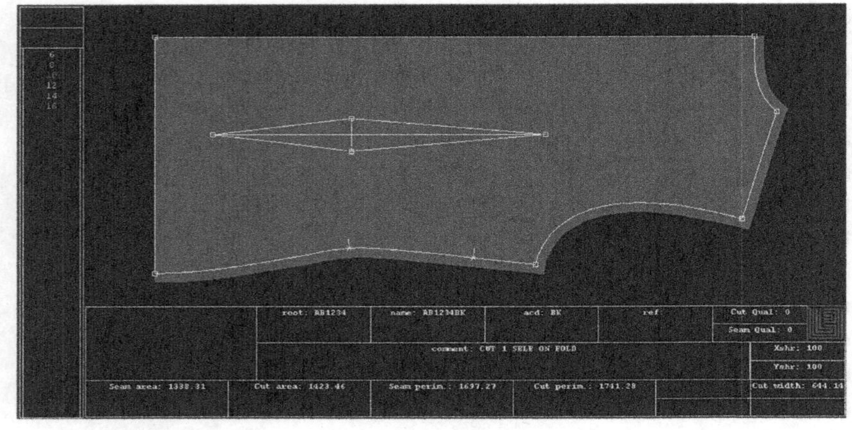

3.12 *Basic block with seam allowances*

Example 2 (see Fig. 3.12 and Fig. 3.13) – In this comment box there are the cutting instructions to 'Cut 1 self on fold'. This is a useful instruction when plotting the pattern out for use manually.

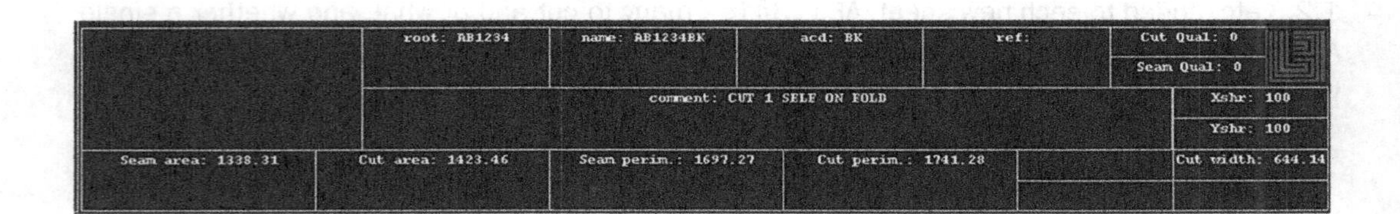

3.13 *Comments box*

Note too, the size range taken from the **model identification sheet** (Fig. 3.12, left-hand border). This will now be applied to each new sheet that is created in the model. As you can see, failure to add it to the model identification sheet, which is the first sheet in any pattern, will mean having to go back and add this information to each subsequent sheet. It is better to do it first when creating the new model.

3.5 Variant

In traditional hand pattern cutting, all the information and instructions for cutting out in cloth are written onto the actual pattern piece. On **CAD** this is no longer the case. The **variant**, which we will look at next, contains all of the cutting information, making annotation onto the pattern piece for computer-performed marker making redundant (Fig. 3.14).

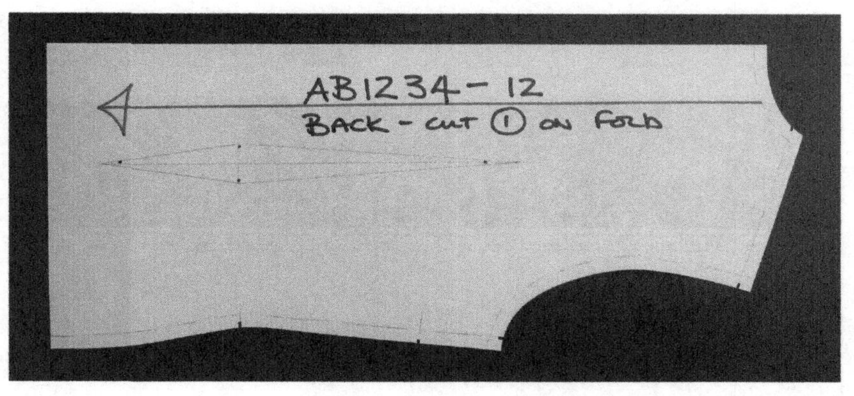

3.14 Card pattern showing hand-written instructions

Once you have made a pattern you will want to create a **variant**, which is the link to **marker making in Diamino**, and contains all of the information relating to how many of each piece you cut, in which fabric and what direction. A pattern can have a number of **variants**, which is the facility to bring together all of the relevant pattern pieces, as one would group card pieces onto a pattern hook. You may use all of them, or have some optional pieces. As an example, a shirt pattern may have a long- and short-sleeve option, while utilising the same body pieces. In this case you may need two **variants**.

For the moment we will look at the text boxes on the **variant sheet** (Fig. 3.15). Although similar to the **model identification sheet**, it has differences.

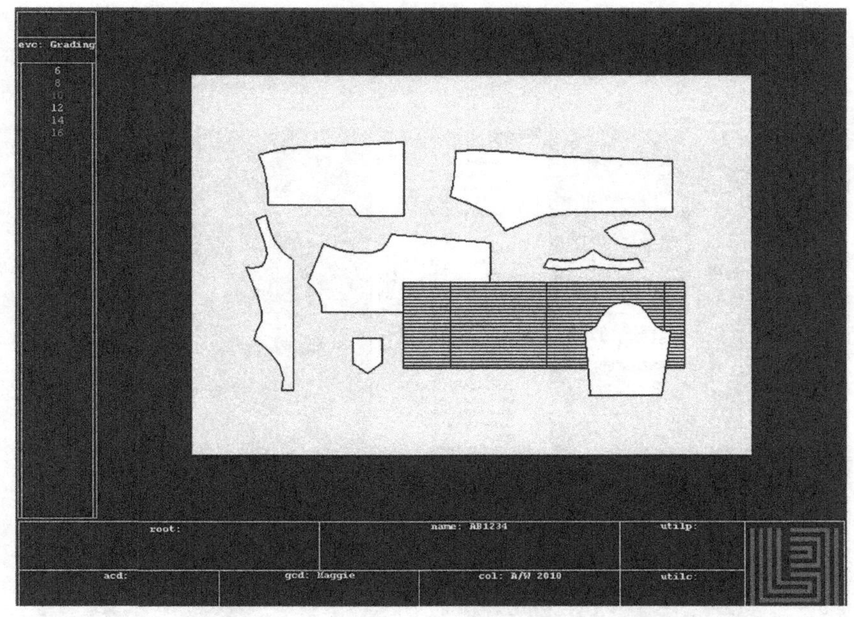

3.15 Variant sheet

The size offer is taken from the **model identification sheet**.

Root: on this page, this is not filled in by default.

Name: here you can name the **variant** according to whatever system you have chosen. It is unique to the variant and so can identify variations within one model. There are a limited number of characters, so coding is useful, for example AB1234SS, adding SS for short sleeve and AB1234LS, adding LS for long sleeve.

gcd: analytical code can have a variety of uses. I use it for the Pattern Cutter's name.

Col: here I add the date the model was created, but you could use it for the season by date, code or words.

Comment: various uses are a description of the model, fabric type or garment variation.

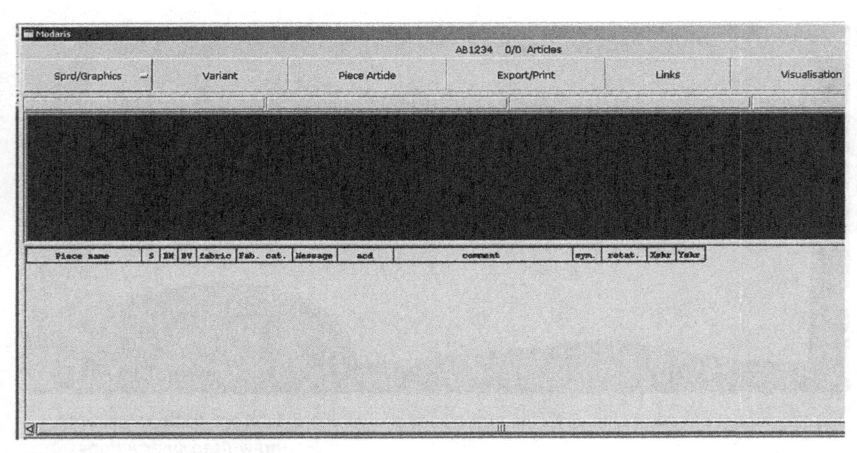

3.16 *Variant spreadsheet empty*

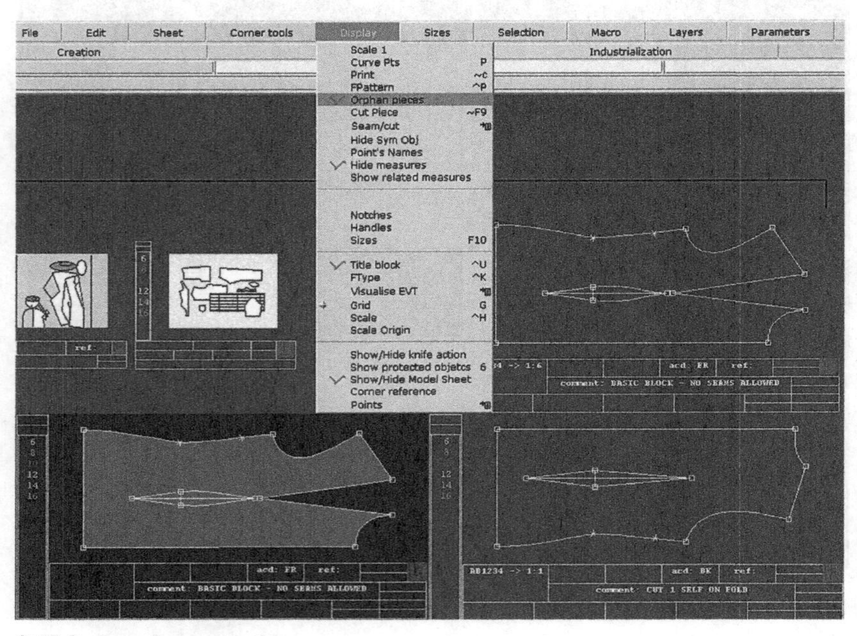

3.17 *Orphan pieces selection from display menu*

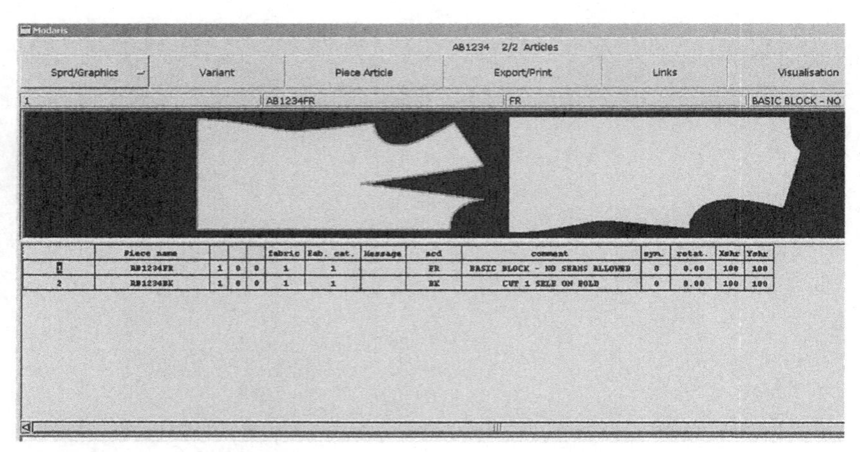

3.18 *Variant spreadsheet added*

3.5.1 *Creating a variant*

To create a **variant** select **F8**, and then **variant** and a new dialogue box will appear in the top left corner of the screen, just like the new model window. Enter the required name of the variant and press **Enter**. To open the **variant**, select **F8** and **variant**, and then click in the centre of the Variant sheet. A new spreadsheet will now open (Fig. 3.16).

Select **F8 – Create piece article**. Click once in the centre of the pattern piece to add to the variant the pieces you require. They will now appear on the spreadsheet. A point to note is that on the main desk, the pattern piece will turn a dark blue, providing you have selected from the drop-down menu **Display – Orphan pieces** (Fig. 3.17).

The information that you have added to each individual sheet will now be displayed in the spreadsheet as follows (Fig. 3.18).

Piece name: by default there will always be a piece name, showing the model name with a unique identifier, either a number, or if you have amended to piece codes, the abbreviation. If you use the same abbreviation more than once, the system will add a sequential number to maintain its uniqueness.

Acd: this will be the same abbreviation that you added to the individual sheet.

Comment: this will be the same as you input on the individual sheet.

It is worth noting here that if you have made a mistake, either in spelling or inputting incorrect information in either of the **acd:** or **comments:** boxes, you can type directly into the spreadsheet box and re-write the information. The rest of the boxes now need filling in manually.

The columns '**S**' '**DH**' and '**DV**' are to indicate how many to cut. Using numbers 0 = none, 1 = one, 2 = two, etc. refers to the number of a single or paired pieces. Note that:

- **S** is for a single piece, so enter 1 for a single piece; 2 will give two pieces, but not paired.
- **DH** is for a double, that is a piece paired on the horizontal axis, so enter 1 for a pair of pieces.
- **DV** is for a double, that is a piece paired on the vertical axis, so enter 1 for a pair of pieces.

Remember to enter 0 into the column that is not required. It is very important that this information is filled in correctly as the marker will only be able to extract the required pieces using this information.

Next comes the **Fabric** column. Typical fabric categories are the 'Self' or outer main cloth, perhaps a contrast outer cloth, a lining and interlinings. Templates for marking pockets or buttons may be needed in card (Fig. 3.19).

Consistent coding for these variations is necessary for **Marker making**. Here's a suggestion:

- **0 – Self-fabric**
- **1 – Fusing**
- **2 – Lining**
- **3 – Card**
- **4 – Contrast fabric**
- **5 – 2nd fusing**

As this column is filled in the icon of the pattern piece will change colour, so that all of the pieces of one fabric type will be colour-coded. It's a bit like manual pattern cutting where different coloured cards were used to differentiate between fabric, linings and templates.

Like many of the other text boxes, the comments box can be used for a variety of purposes. I use it to supplement the fabric's information by stating in words what the cloth is – self, or satin, or poplin – the fusing quality, or non-fuse, jersey or wovens. This helps me to ensure that the other boxes, S, DV and DH are correct.

The **variant** is a vital communication tool and time must be taken to ensure that all of the pattern pieces are included, cutting instructions correct and of the right fabric. Take your time.

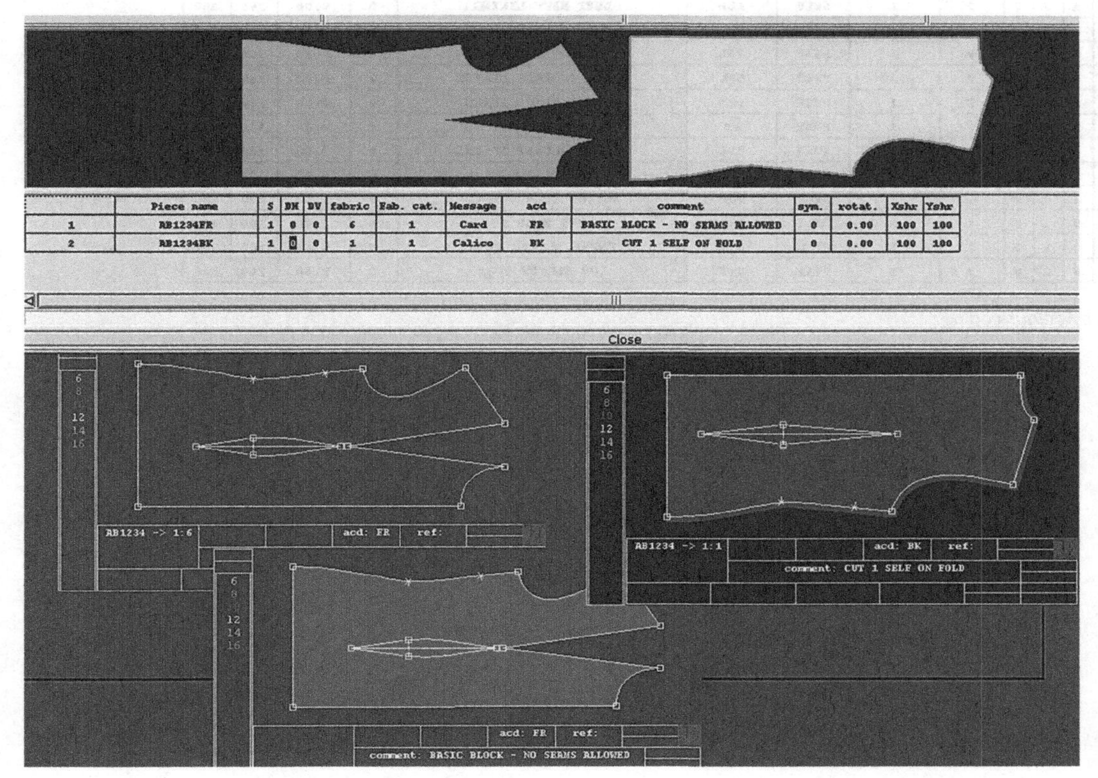

3.19 Screen showing fabric types by colour coding

3.6 Industry example

We have looked at a very small number of pattern pieces while discussing the various text boxes, so Fig. 3.20 shows a full pattern of a simple jacket on screen showing the different cloth types that will be used and the accompanying variant spreadsheet.

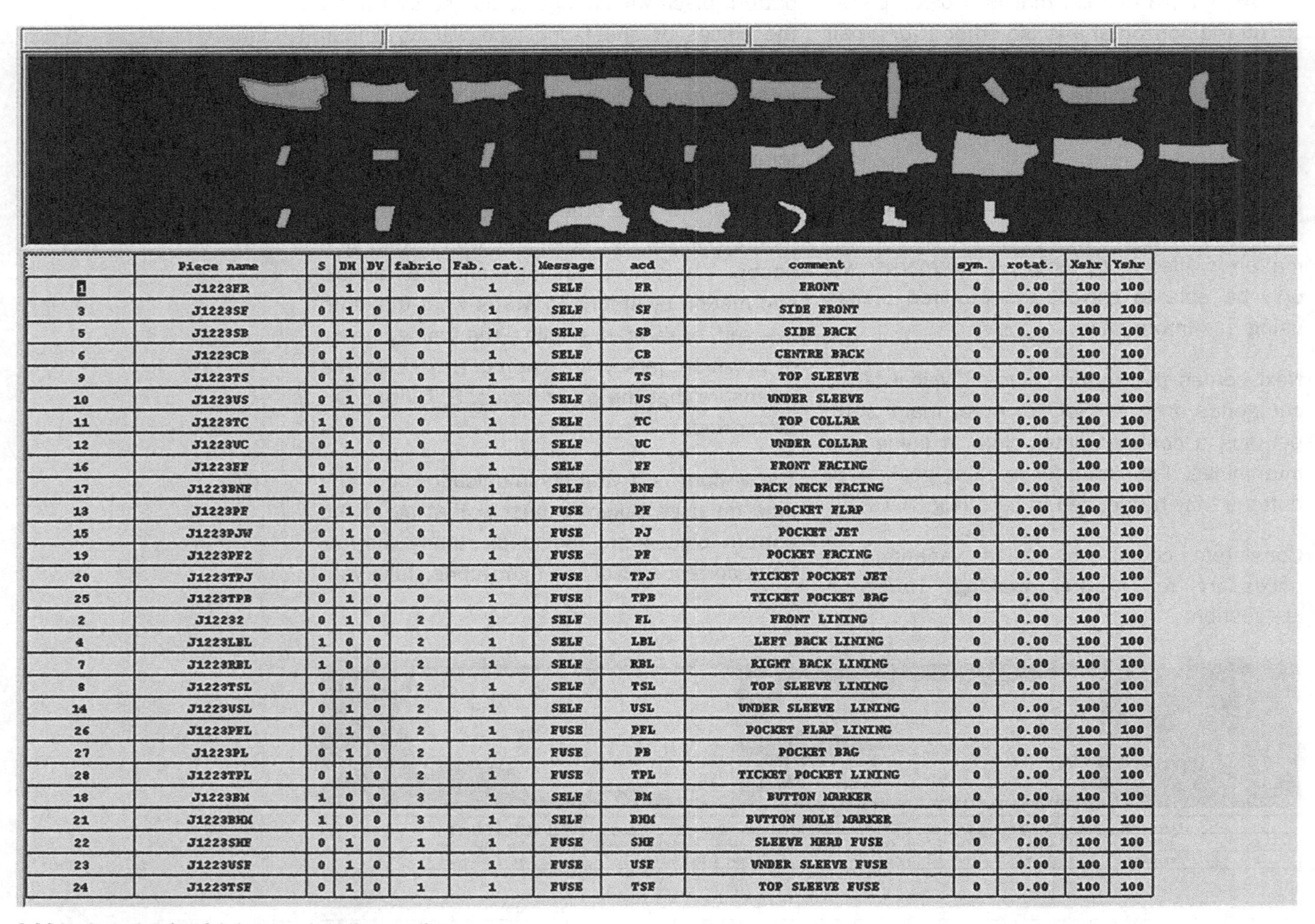

	Piece name	S	DH	DV	fabric	Fab. cat.	Message	acd	comment	sym.	rotat.	Xshr	Yshr
1	J1223FR	0	1	0	0	1	SELF	FR	FRONT	0	0.00	100	100
3	J1223SF	0	1	0	0	1	SELF	SF	SIDE FRONT	0	0.00	100	100
5	J1223SB	0	1	0	0	1	SELF	SB	SIDE BACK	0	0.00	100	100
6	J1223CB	0	1	0	0	1	SELF	CB	CENTRE BACK	0	0.00	100	100
9	J1223TS	0	1	0	0	1	SELF	TS	TOP SLEEVE	0	0.00	100	100
10	J1223US	0	1	0	0	1	SELF	US	UNDER SLEEVE	0	0.00	100	100
11	J1223TC	1	0	0	0	1	SELF	TC	TOP COLLAR	0	0.00	100	100
12	J1223UC	0	1	0	0	1	SELF	UC	UNDER COLLAR	0	0.00	100	100
16	J1223FF	0	1	0	0	1	SELF	FF	FRONT FACING	0	0.00	100	100
17	J1223BNF	1	0	0	0	1	SELF	BNF	BACK NECK FACING	0	0.00	100	100
13	J1223PF	0	1	0	0	1	FUSE	PF	POCKET FLAP	0	0.00	100	100
15	J1223PJW	0	1	0	0	1	FUSE	PJ	POCKET JET	0	0.00	100	100
19	J1223PF2	0	1	0	0	1	FUSE	PF	POCKET FACING	0	0.00	100	100
20	J1223TPJ	0	1	0	0	1	FUSE	TPJ	TICKET POCKET JET	0	0.00	100	100
25	J1223TPB	0	1	0	0	1	FUSE	TPB	TICKET POCKET BAG	0	0.00	100	100
2	J12232	0	1	0	2	1	SELF	FL	FRONT LINING	0	0.00	100	100
4	J1223LBL	1	0	0	2	1	SELF	LBL	LEFT BACK LINING	0	0.00	100	100
7	J1223RBL	1	0	0	2	1	SELF	RBL	RIGHT BACK LINING	0	0.00	100	100
8	J1223TSL	0	1	0	2	1	SELF	TSL	TOP SLEEVE LINING	0	0.00	100	100
14	J1223USL	0	1	0	2	1	SELF	USL	UNDER SLEEVE LINING	0	0.00	100	100
26	J1223PFL	0	1	0	2	1	FUSE	PFL	POCKET FLAP LINING	0	0.00	100	100
27	J1223PL	0	1	0	2	1	FUSE	PB	POCKET BAG	0	0.00	100	100
28	J1223TPL	0	1	0	2	1	FUSE	TPL	TICKET POCKET LINING	0	0.00	100	100
18	J1223BM	1	0	0	3	1	SELF	BM	BUTTON MARKER	0	0.00	100	100
21	J1223BHM	1	0	0	3	1	SELF	BHM	BUTTON HOLE MARKER	0	0.00	100	100
22	J1223SHF	0	1	0	1	1	FUSE	SHF	SLEEVE HEAD FUSE	0	0.00	100	100
23	J1223USF	0	1	0	1	1	FUSE	USF	UNDER SLEEVE FUSE	0	0.00	100	100
24	J1223TSF	0	1	0	1	1	FUSE	TSF	TOP SLEEVE FUSE	0	0.00	100	100

3.20 Variant showing fabric types by colour coding

The following story illustrates the importance of accurate information

A factory, while developing a t-shirt design, plotted off the pattern and gave it to a Sample Cutter who cut the paper pattern pieces out, laid them onto cloth and cut out the garment, as he saw fit and according to his experience. There was no written information on the paper pattern because the factory would eventually use **Diamino** for its marker making for production garments. During the sample development the short sleeve was cut in a woven cloth and was cut on the straight grain, giving a neat, slim-fitting short sleeve. In fact on the computer screen the sleeve was cut on the bias, but the grain line had not been plotted out clearly enough for the cutter to see that it was a bias cut sleeve. In production the real orientation of the sleeve was used and the garments were made with a fluted sleeve not a slim sleeve as the samples. The production run of 2000 garments was rejected.

Chapter 4

Digitising a clothing pattern on Lectra Modaris pattern cutting software

Abstract: Digitising is a way to transfer a paper pattern into the computer. This chapter will explain the hardware, how to use it and how to avoid problems. The options regarding the preparation of the paper pattern are also discussed, followed by a step-by-step guide to the process of digitisation.

Key words: digitising, handset button notes, paper pattern preparation, pattern numbering, piece names, Lectra Modaris.

4.1 Digitising board

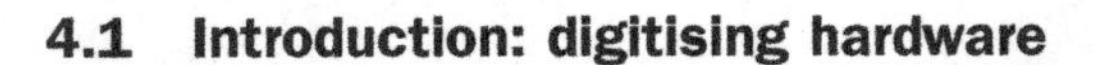

4.1 Introduction: digitising hardware

Digitising is a way of taking a paper pattern and transferring it to the computer system, in other words making it digital. This requires an interactive digitising table or board (Fig. 4.1) activated by the use of a digitising mouse (Fig. 4.2), sometimes referred to as the keypad or handset. This may be a cordless version or joined to the digitiser by a lead.

When the handset is held against the digitising table and operated by its command keys, the pattern shape is recreated within Modaris. At the top of the handset is a transparent section containing what are referred to as crossed hairs. When digitising, the place of intersection of the crossed hairs is the point that will be transmitted to the computer so it makes sense to be as accurate as possible while digitising.

4.2 Digitising handset or mouse

The functions assigned to the 16 digitising keys on the handset can be used singly or in combination to perform complex operations. To begin with I would suggest keeping to the simple but perfectly adequate procedure explained in this chapter. Digitising can also be done in stages, entering one set of pattern pieces and saving. When you are ready, digitise the next set of pieces. Pattern pieces can also be added to an existing pattern at any time.

Open the required model, create a new sheet and follow the digitising guide. I suggest you photocopy and attach the keypad button menu to the digitising board for easy reference until the procedure is fully familiar (see Fig. 4.3). Please note Key A is normally used to define the horizontal axis.

1 Characteristic point	**2** Corner point	**3** Adds an internal mark or drill hole (**Relative point** in **F1**)	**C** Curve point
4 Pattern hook hole (A nod to card patterns requiring a hole)	**5** Used with 3, 6 & 8 to vary the drill hole, notch or line type	**6** Adds a notch to the point	**D** Cancels the last action
7 Use to define the beginning of an internal line	**8** Use to create a reference line	**9** Use to orientate a notch	**E** Enables the naming of the last point
A Grain line Validation key Used for grading mode	**0** New sheet	**B** Creates a notch on a slider	**F** Finishes the last action, i.e. an internal line FF completes digitising

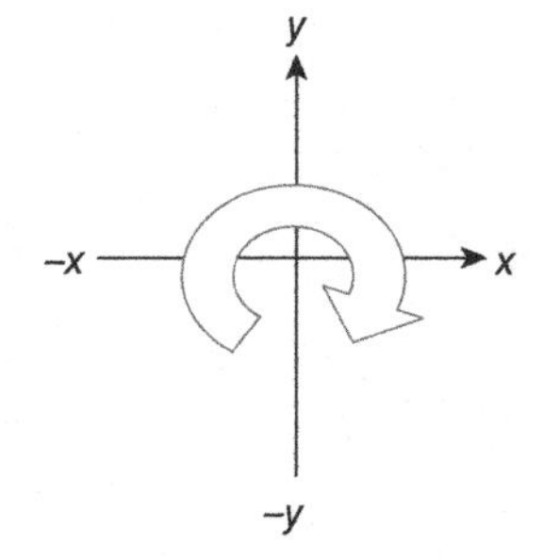

4.3 Handset button menu

4.2 Digitising with or without seam allowances

Pattern cutting skills are different to computer skills and a person may be skilled in the one and not the other. This will influence the decision to digitise the pattern with seam allowances already added or without seam allowances as a net pattern. Some pattern rooms use paper patterns for the initial samples, and make any amendments required manually until the master paper pattern is approved. The perfected pattern is then digitised onto the system. Other pattern rooms work directly on screen, creating and amending patterns on Modaris. There are advantages and disadvantages for each of these methods.

4.2.1 Digitising with seam allowances

Advantages of digitising with seam allowances are as indicated below.

- Any person with the appropriate computer skills and a little training can digitise the pattern.
- A Pattern Grader, Marker Maker or a Junior Pattern Cutter without advanced pattern cutting skills can digitise the pattern.
- There is no need to add seam allowances or adjust corners as the pattern will have these already, although some notches will still need to be added.

Disadvantages of digitising with seam allowances are outlined below.

- A fully finished pattern is required.
- Pattern amendments and changes in seam allowances are more difficult to execute.
- Style development from an existing pattern is more complex.

Figure 4.4 shows the method of digitising a pattern that has seams already added. Note the corner shapes with two 2's used to define the seam width.

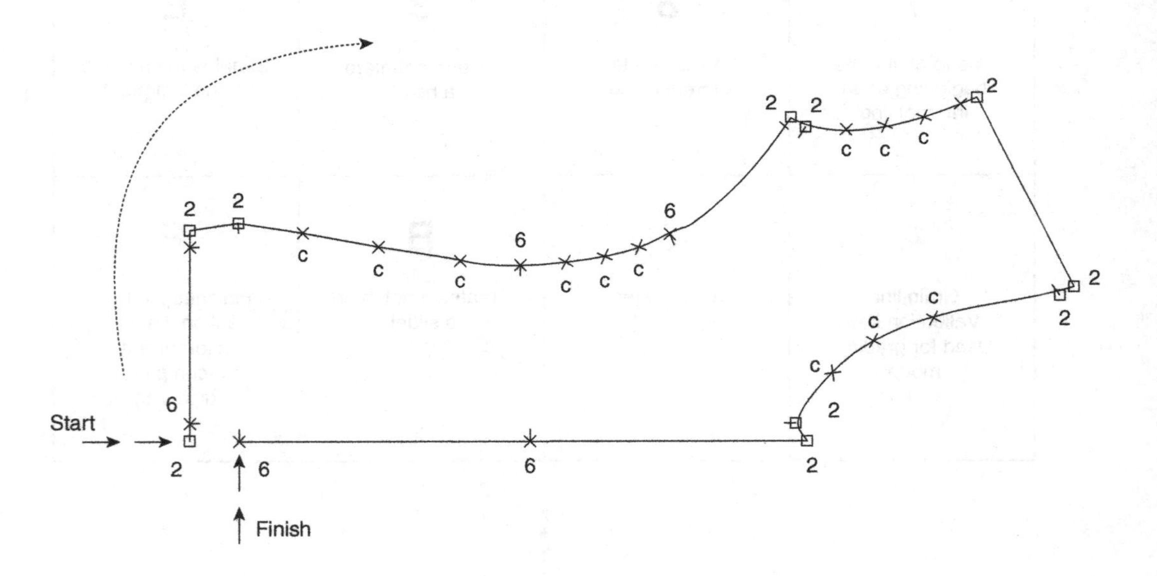

4.4 Digitising a pattern with seam allowances illustrating button key commands

4.2.2 *Digitising without seam allowances*

Advantages of digitising without seam allowances are outlined below.

- Only the basic draft is required, which is then finished on screen.
- Work is with the true sew-line.
- Seam allowances can be altered very easily.
- Alterations and new pattern pieces can be made on the true sew line.
- Style development is straightforward.

Disadvantages of digitising without seam allowances are indicated below.

- Pattern cutting skills are required.
- Advanced Modaris skills are required.

Figure 4.5 shows the method of digitising a pattern piece that is without seam allowances. Note that the corners have only one 2 as seam allowance will be added in Modaris.

Figure 4.6 shows the two finished shapes on Modaris. Note the position of the white line relative to the seam allowance. See Chapter 7, Section 7.2.8 F4 – exchange.

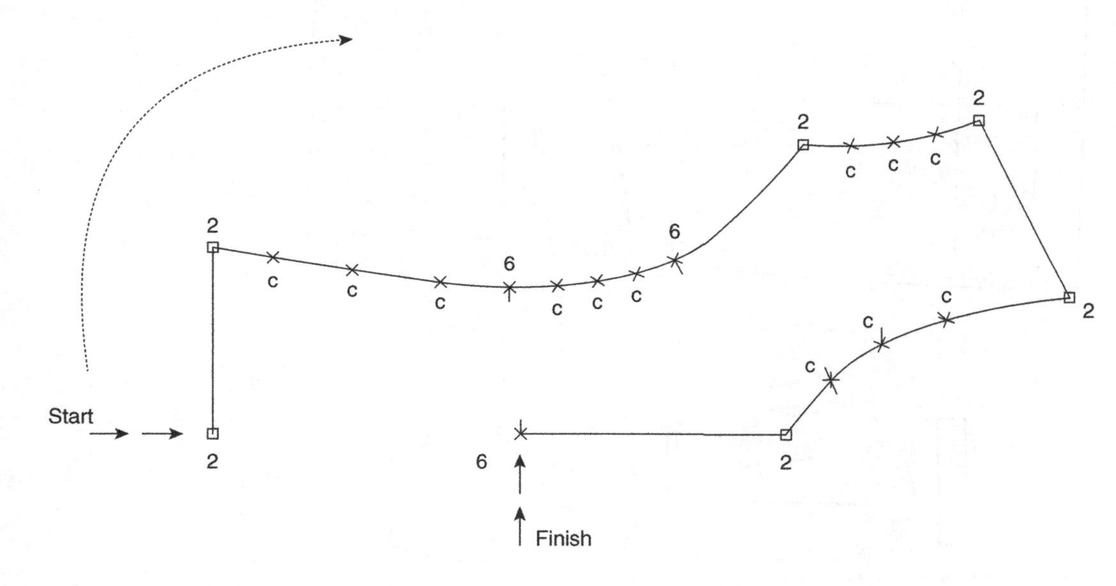

4.5 Digitising a pattern without seam allowances illustrating button key commands

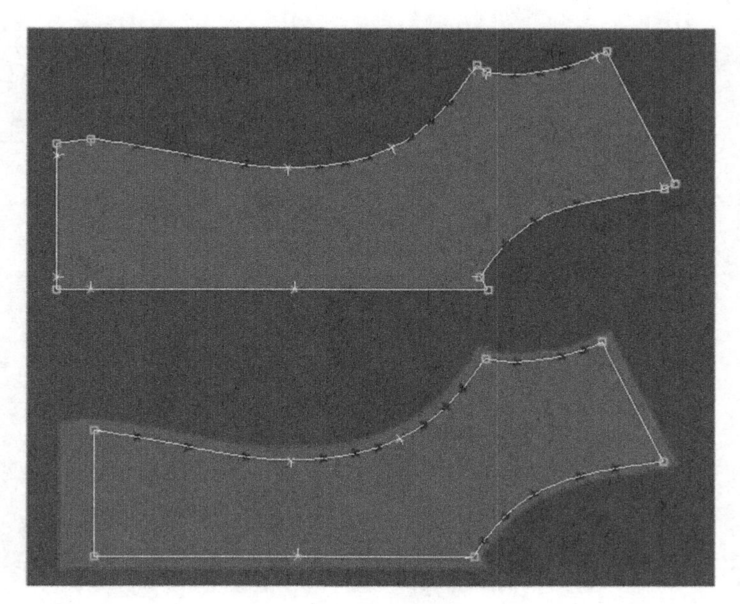

4.6 The same pattern pieces on Modaris

4.3 Pattern preparation

Good preparation of the paper pattern prior to digitisation is essential, regardless of whether it is with or without seam allowances, as the pattern on screen will only be as good as the original combined with the care of the digitiser (see Fig. 4.7).

Check that:

- the seam lines are a good shape;
- all of the seam lines which join together match;
- seam allowances and corners are correct;
- the grain lines are drawn in correctly;
- all of the notches and balance marks are drawn in;
- any internal lines, darts or drill holes are clearly marked.

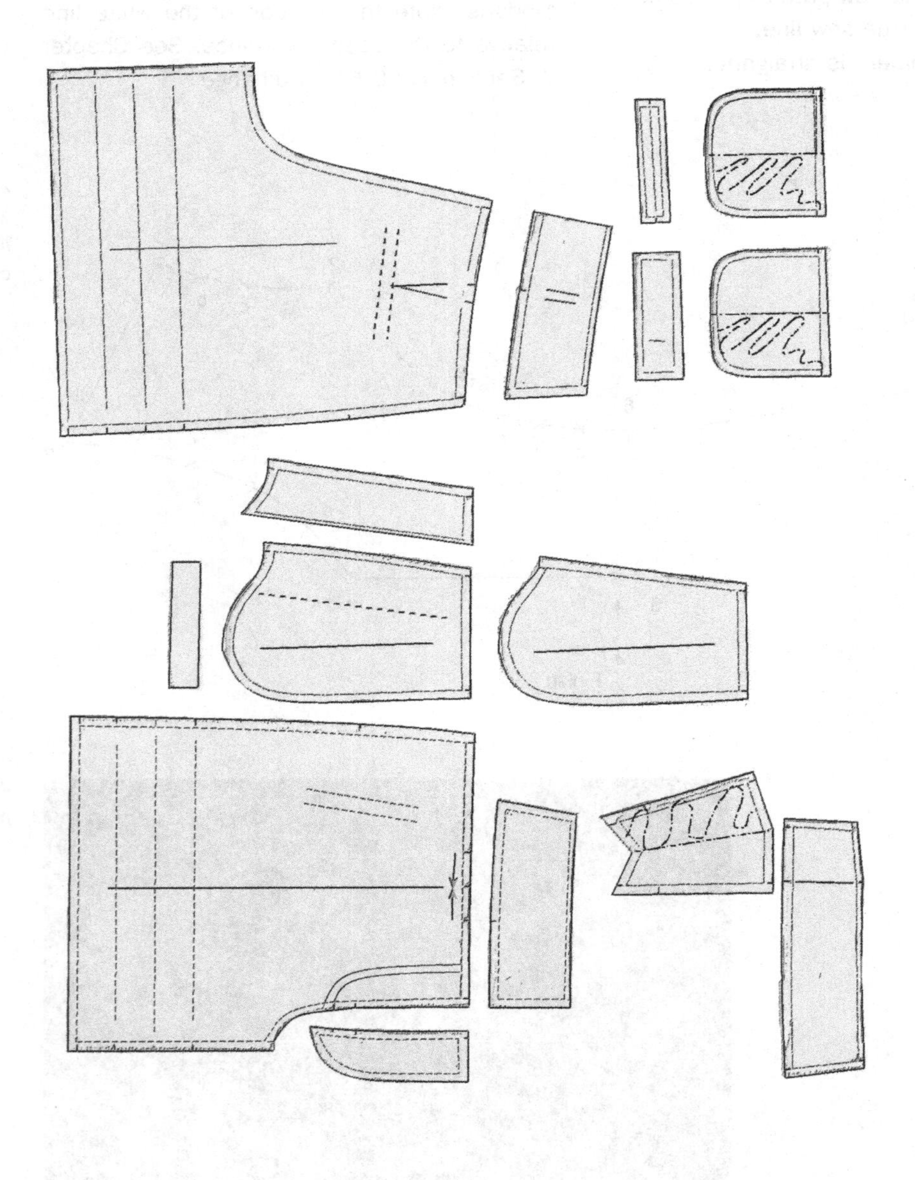

4.7 Complete paper pattern needed for digitising with seams allowed

Additionally if you are adding seam allowances afterwards on Modaris:

- pockets, tabs and other small parts are drawn directly onto the required pattern piece; there is no need to make a separate piece;
- facing positions need only be drawn onto the draft pattern;
- many derived pieces, such as lining and interlining pieces, need not be made at this stage.

See Figs 4.8 and 4.9. Figure 4.9 shows the finished back pattern piece on Modaris with seam allowance now added.

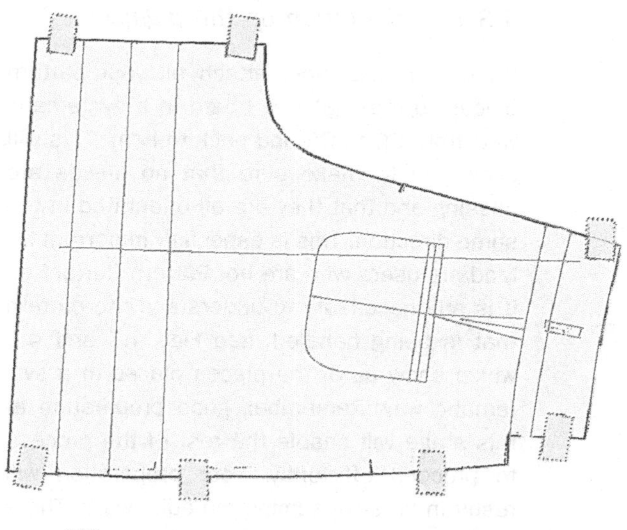

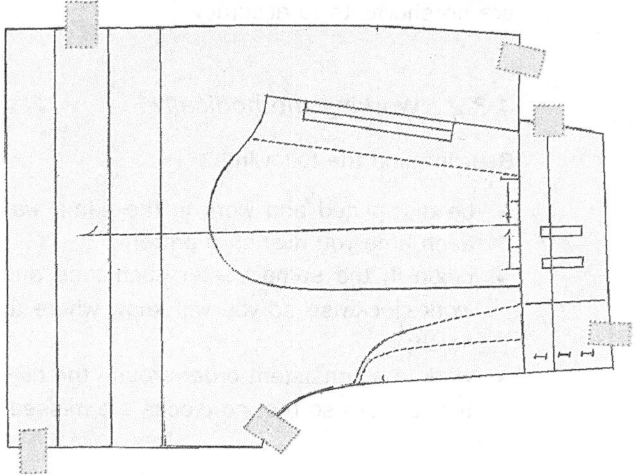

4.8 The same style paper pattern in draft form prepared for digitising

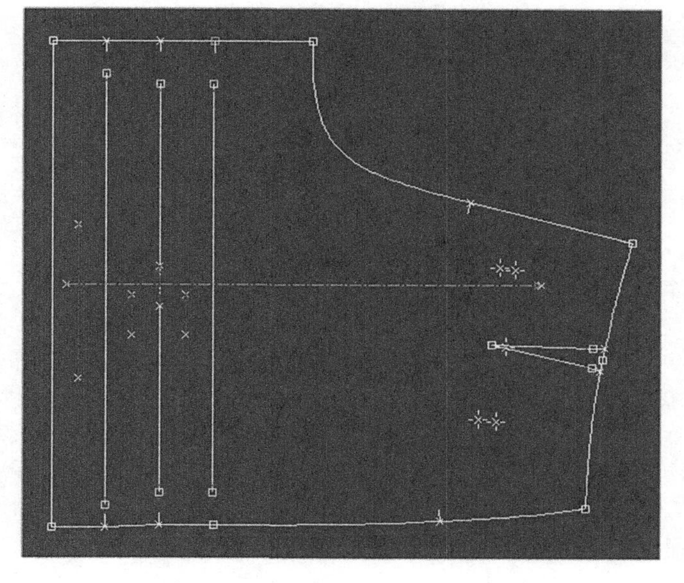

4.9 The back leg from the draft pattern completed on Modaris

4.3.1 *Orientation on the digitiser*

Using masking tape, attach all your pattern pieces to the digitising board in a systematic way: from CB to CF, and neck to hem. This will allow you to make sure that no pieces are missing and that they are all orientated in the same direction. This is especially important for Modaris users who are not Pattern Cutters as it is still necessary to understand the pattern that is being handled, see Figs 4.7 and 4.8 which show all of the pieces placed in a systematic way. Remember, good preparation at this stage will enable the rest of the process to proceed efficiently. Poor preparation will result in time-consuming remedial work. There are no shortcuts to accuracy.

4.3.2 *Working methodically*

Bear in mind the following:

- be disciplined and work in the same way each time you digitise a pattern;
- begin in the same corner each time and work clockwise so you will know where to finish;
- work in a consistent order around the digitising board so that no pieces are missed.

See Figs 4.4 and 4.5 for reference. In a busy pattern room it is inevitable that you will be interrupted but, if you know where you have begun and the order in which you work, it will be easy to pick up your work without too much trouble.

4.3.3 *Pattern identification*

On a computer your patterns will be filed in numerical or alphabetical order by default, organised by the computer and only identifiable by the number or short code that you decide to use. Trying to remember every pattern shape in your library just by a code will be demanding. A few key patterns can be referred to very easily but, once your library has hundreds or perhaps thousands of patterns, it will become impossible to remember them all.

Keep a visual record using small-scale sketches placed next to the allocated pattern number in a file which will help identify patterns for future use. How you number is entirely personal, but there has to be a system in which you store patterns. Prior to digitising be prepared with the sketch and pattern number and any associated paperwork required.

4.3.4 *Creating a new file and labelling the model identification sheet*

Use the following set of instructions.

- Open **Modaris**.
- Select **File – New** and a dialogue box will appear in the top left corner of the screen. Enter a new unique number or name (see Fig. 4.10).
- Press **Enter** and then **Save**. It is important to note that there should be no prompt window telling you that this file already exists. This is the only time that there will be no prompt window appearing, and shows that this number is indeed a new one that does not already exist.
- If a window appears saying '**This file already exists**' (as in Fig. 4.11) do not continue, as you will overwrite an already existing file. In this case select **Cancel/ Abort**, allocate a new number and start again.
- Select **Display** – **Title blocks**. A new margin of the title page that allows for text will now appear.
- Press **J** or **j** on the keyboard to resize page.
- Go to **Edit** – **edit**, click in the relevant title block cell (text box) and type in the pattern naming information required.

See Chapter 3 on title blocks for a fuller explanation of their use.

4.10 New file window

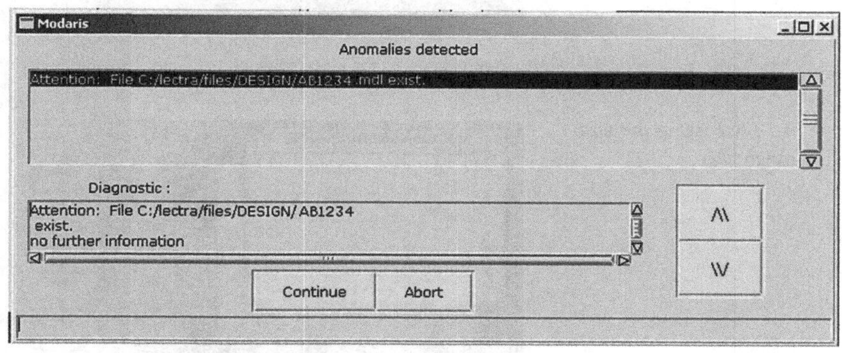

4.11 Prompt window – file already exists

4.12 F7 Imp EVT

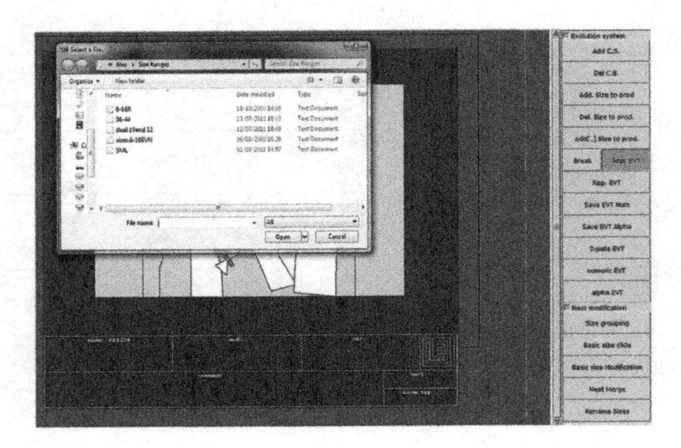

4.13 Accessing the size range files

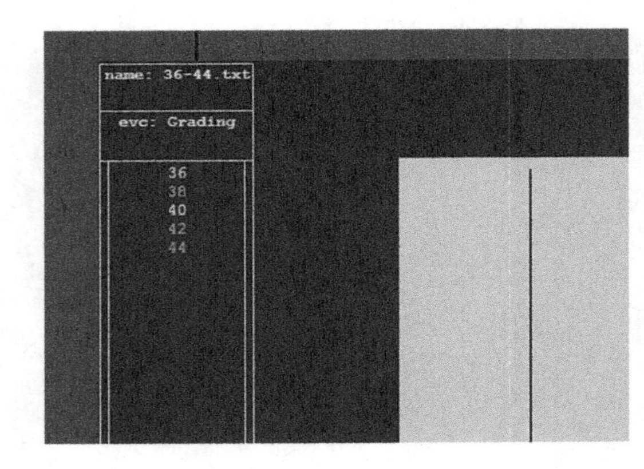

4.14 Size range text box showing the range of sizes

4.3.5 Adding the size range

Use the following set of instructions.

- Select **F7 – Imp EVT** to gain access to the background size range file (see Fig. 4.12).
- Double click in the centre of the title page to open the size range file (see Fig. 4.13).
- Select and double click on the size range that you wish to use. The same range will appear down the left-hand side of the title page (see Fig. 4.14). Note: your pattern will not plot out if it does not have a size range to relate to.
- Now **Save** and you are ready to digitise your pattern.

When opening the size range file, Lectra recommends displaying the file extensions at the same time. See Chapter 11 for how to create a size range file.

4.3.6 Saving work and folder options

Save will always go into the default folder. Lectra have set this as the **Design** folder for the default, but this can be customised as required. If you want to **Save** into a different folder, e.g. **blocks,** or a specific **client** folder then use **Save as** and select the required folder. You will need to do this each time you save if the pattern is not going into the default folder.

4.3.7 Adding to an existing pattern

Digitising is not only confined to new patterns and can take place at any time. To add a new pattern piece to an already digitised pattern, open the required model and create a new sheet. The initial preparation described above will already have been done, so simply continue as below.

4.4 The digitising process

4.4.1 Activating the digitising function

Having created a new file, added a size range and saved the file for the first time, go to **Sheet – new sheet** and a new sheet will appear on screen. Use **Home** to bring it to full screen.

Select **F1** – **Digit** and notice on the computer screen that an information box has appeared in the top left of the screen saying '**2 points for horizontal axis**'. This confirms that the digitising function has been activated, and is purely a help box.

With your mouse, place the cursor so that it is in the centre of the sheet on the screen. Leave it there, put down the mouse and pick up the digitising handset.

4.4.2 Activating the digitising board

In one corner of the digitising table, usually near the on/off switch, there is a small flickering light. When the handset is held against the board, this will become a steady light indicating that the board can read the handset and digitising can begin.

The following points may be of use if the light continues to flicker.

- Hold the handset away from the board, press any button a couple of times and try again.
- Turn the digitising board switch off for a few seconds and then back on.
- At the lower right of the main computer screen is a small icon for the digitiser. Double click on the icon and refresh the application, making sure that the correct serial port has been selected, and select OK. Then try again (see Fig. 4.15).
- The handset may need a new battery.
- Rarely, the handset may need replacing.

4.4.3 Defining the grain line

- To start, hold the handset against the board on the first piece, and use **A** + **A** to define the straight grain line.
- Press **A** at the beginning of the grain line and **A** at the end. This will orientate the piece on screen, with the grain line on the x-axis.

Ensure consistency when digitising grain lines. Always move left to right or right to left, so that all pieces are orientated in the same direction.

On small pieces this action may not register because there is insufficient distance between the beginning and end of the line. In this case, extend the grain line outside of the piece and this will enable digitisation (see Fig. 4.16).

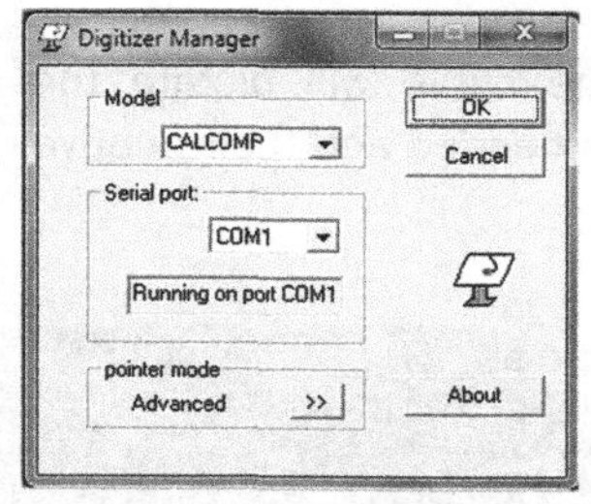

4.15 Digitising manager refresh box

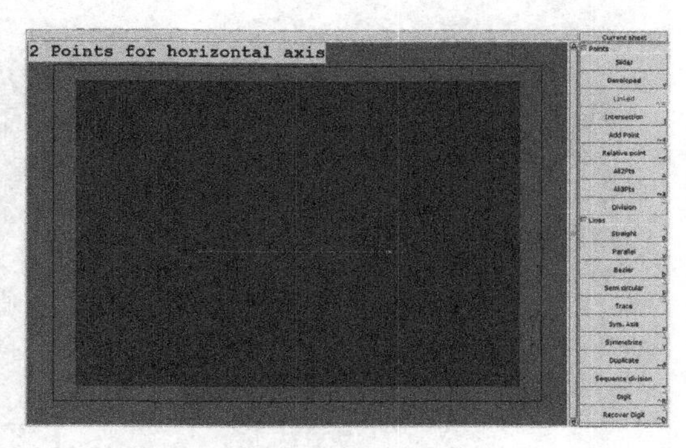

*4.16 Digitising the grain line, **A – A***

4.4.4 *Digitising the pattern shape*

Begin the outline of the pattern shape at a corner using **2** and continue around the piece in a clockwise direction using **2** for corner points, **C** for curved points and **6** for notches (see Fig. 4.17).

Using only these three keys is the simplest method of digitising, and I would recommend becoming very familiar with this process before using the more complex features of the handset functions.

Notes for **2** points:

- use at the beginning and end of a straight line,
- use for sharp definition, e.g. seam allowance corners.

Notes for **C** points:

- always use an odd number and as few as possible.
- the shape of the curve will dictate the number of points needed and their relative position to each other; the more extreme the curve, the closer the points.

Notes for **6** point:

- creates a characteristic point (this means the point is sharp rather than curved) and adds a notch; can be used anywhere a notch is required.

Notes for **D** key:

- while you are working, the points and lines created will appear on the computer screen, allowing you to check on progress; if a key has been struck in error, **D** will delete that last point and take you back a step.

Stop digitising one point before reaching your first point, as there is no need to complete the sequence. Lectra recommends closing the shape using F before proceeding with internal lines or drill holes. Before leaving this piece add any internal lines or drill marks.

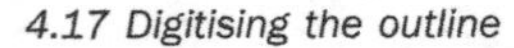
4.17 Digitising the outline

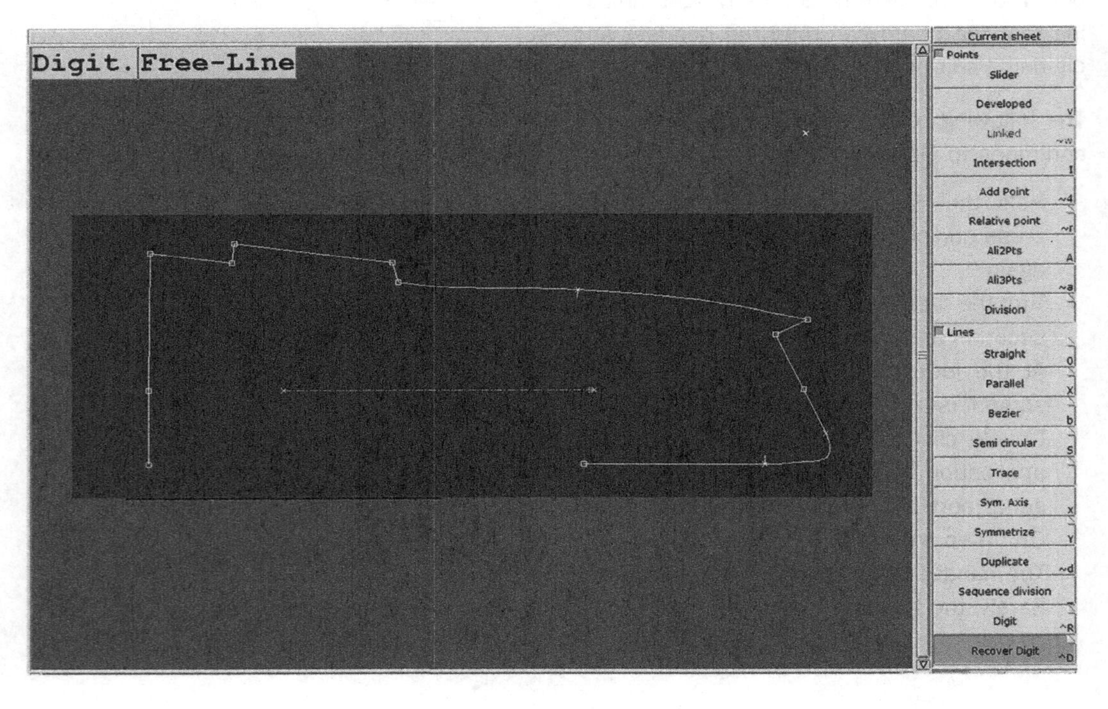

4.4.5 *Adding internal lines and drill holes*

- To add an internal line, the sequence starts with **7**, then use **1** or **C** as needed and end with a single **F**. Do NOT try to add an internal line attached to the perimeter line. Click a little inside the line using **7** to begin the line. Continue as before using **1** and **C** with **F** to finish, again a little away from the perimeter line.
- You can add further internal lines to the same piece as required.
- An example of the sequence is **7–1–C–C–C–C–C–C–C–1–F** (see Fig. 4.18).
- Add drill hole marks for button positions or dart ends using **3**.

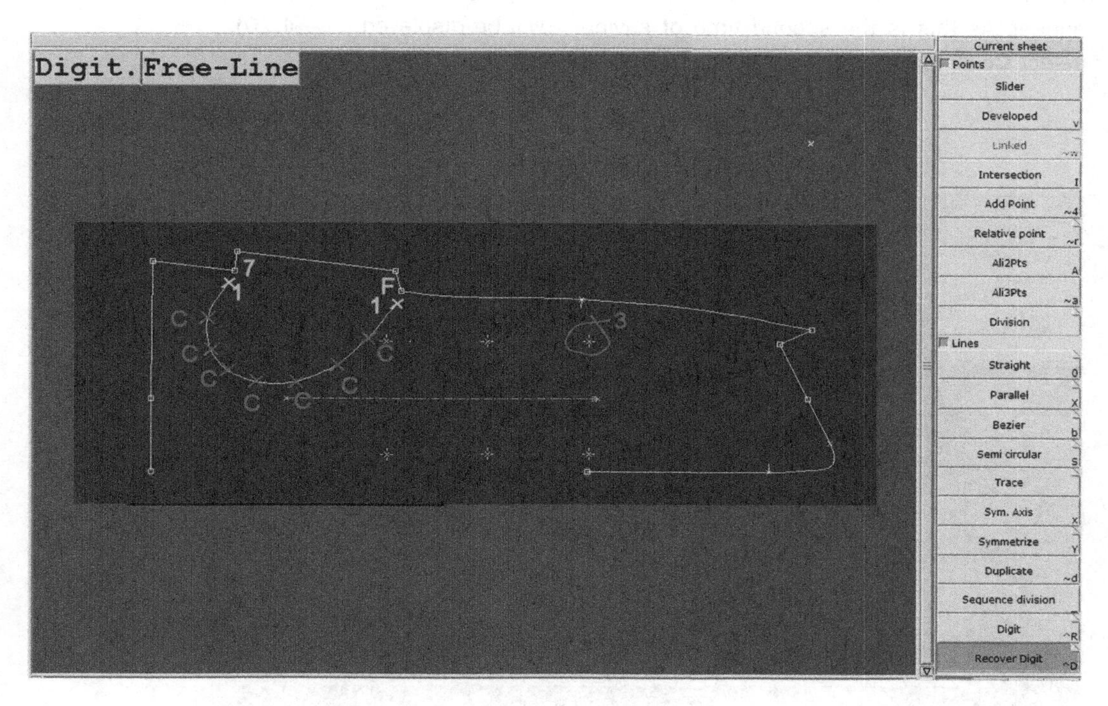

4.18 Digitising internal lines and points

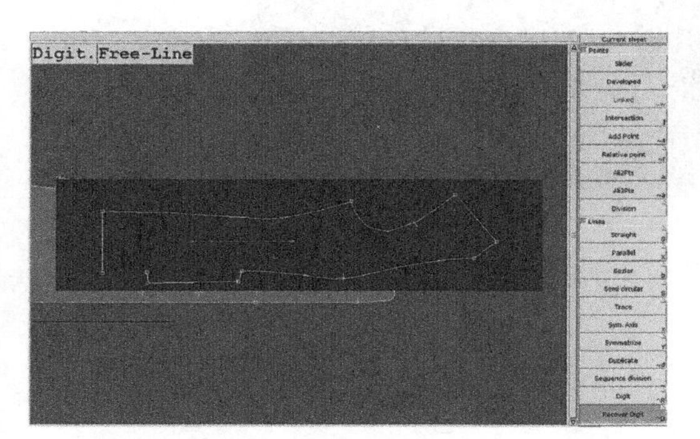

4.19 Digitising subsequent pieces

4.4.6 *Creating a new sheet*

When you have completed work on a particular piece, move on to the next by using **0** to create a new sheet. This action will complete the previous sheet, joining the first and last points, and create a new sheet to work on (see Fig. 4.19).

4.4.7 Finishing and exiting digitising mode

When all the pieces are complete, use **FF** to finish the digitising function. On screen now will be a jumble of digitised sheets probably overlapping each other. By pressing **J/j** on the main keyboard a couple of times they will re-arrange and all be visible on screen. **Save** the pattern and this time a prompt window will appear. As this is the second time of saving, select **Continue**.

4.4.8 Flat pattern pieces

Digitising also has a set of associated parameters, i.e. some display options. It is possible to digitise so that the grey sheet containing the flat patten layer (Fig. 4.19) is not displayed after the completion of the digitising session. Click on the corner of the **F1 – Digit** menu to open the associated parameters menu and tick **No flat pattern**. Now only the blue sheets will be displayed (Fig. 4.20).

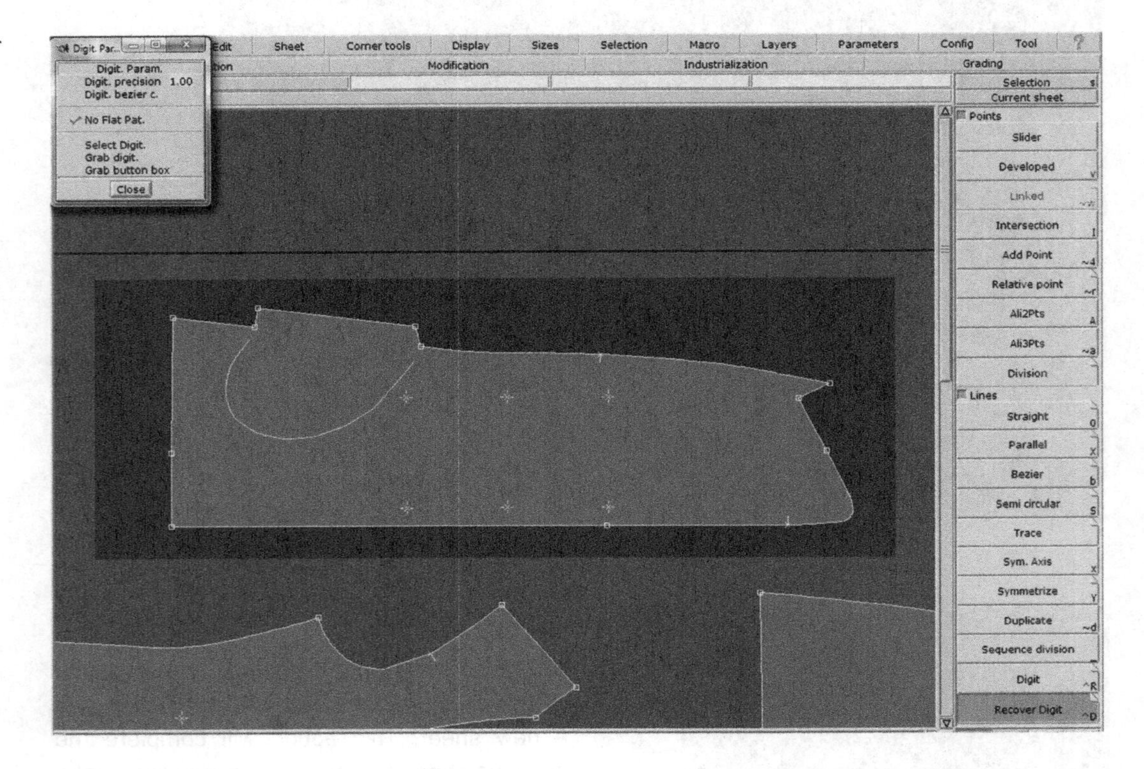

4.20 Pattern piece after completing digitising.

4.5 Handset button notes

0 – Creates a new sheet.
1 – Adds a characteristic point (mostly used for internal line creation).
2 – Adds a corner point (two of these are required on each piece).
3 – Adds an internal mark, the **relative point** in **F1**, used to mark drill holes or button placements.
4 – Adds a pattern hook hole (a nod to card patterns which had a hole punched for the hook).
5 – Used with 3 or 6 and 8, this button will vary the type.
6 – Adds a notch on a characteristic point **1**.
7 – Use to define the beginning of an internal line.
8 – Use to create a reference line.
9 – Use to orientate to notch.

A – used to determine the horizontal axis.
A, B and E – Used in combination with other functions.
C – Adds a curve point.
D – Cancels the last action.
F – Finishes the last action, i.e. completes an internal line.
FF – Finishes the digitising session.

4.6 Digitising problems

4.6.1 Failure to exit digitising mode

Occasionally I press the wrong keys and get stuck in a cycle that will not let me finish digitising. This will be identifiable by a repeated dull beep each time the **F** button is struck. If this happens, put down the handset and go to the keyboard. **FF** works just as well on the keyboard. Ensure that capital (**Shift**) is selected and strike **F**, two times and the digitising mode should end.

4.6.2 Digit and recover digit

At the start of digitising, **F1 – Digit** is selected. As soon as the actual digitising process begins, **recover digit** will be highlighted. This is normal, so do not try to alter or the process will be interrupted and then difficult to complete.

To cancel **digit**, before any digitising has taken place, right click on the main mouse. The function will end.

To cancel **recover digit**, strike **F** two times on the keyboard. If the function does not end, check that the **Caps lock** is on. Lower case **f** will not work. Try again.

4.6.3 Small errors and deleting commands

Remember that if you make a mistake while digitising, striking **D** will undo the last command and take you back a step. This will work for as many steps as you need.

4.6.4 Big errors and restarting

Sometimes, for a variety of reasons, digitising does not proceed successfully. In this case, use **FF** to complete the function and return to the computer screen. Assess the progress and delete any inaccurate sheets. Save your work. Create a new sheet and start digitising again at the appropriate place.

Digitising an entire pattern does not have to be completed in one go. The main pieces can be entered and then saved. The next set of pieces can then be entered and saved, and so on. This is a suitable method for a complex pattern which is either composed of very large pieces or very many pieces.

Chapter 5
Plotting using JustPrint

Abstract: JustPrint is the Lectra software program used for selecting a Modaris file and sending it to a plotter or printer. This chapter will explain the settings required for the software as well as how to select and send a model or selected pieces to plot.

Key words: JustPrint, plotter, geometry, text, text label, plot pieces, plot model, Lectra Modaris.

5.1 Introduction to JustPrint V2R2

JustPrint (Fig. 5.1) is the software used to plot out patterns. In essence it is very straightforward once the various settings for the program have been chosen and saved. This is explained in Section 5.2. JustPrint can send files to a variety of plotters and printers. You will need to refer to your IT department or a Lectra engineer to set out the choices available to you.

5.1 *JustPrint icon*

The guidance from Section 5.4 onwards will explain the settings. Once the basic principles are understood you can personalise them as you need. It may be helpful to refer to Chapter 3, Section 3.2 Title blocks, as it will now become clear why it is necessary that the various text boxes are annotated accurately, the correct axis lines used and the variant filled in with care.

5.1.1 Information required in order to plot a model

Let's assume that all the JustPrint option settings are in place. There are four aspects of the model to define before sending it to plot:

- **model number**, chosen from the file list within your chosen folder;
- **variant option**, chosen via the search button (Fig. 5.2);
- **fabric type**, chosen via the search button, (Fig. 5.2);
- **size**, using the default base size setting or by selecting another size via the search button (Fig. 5.2).

5.2 *Search button*

5.2 How to plot out a model using JustPrint

- Launch **JustPrint** and select **Add . . .** in the top field in the right-hand column (Fig. 5.3).
- This in turn will launch a second window called **add elements to the batch** (Fig. 5.4).

5.2.1 *Add model file*

- Click once on the model of your choice and it will become highlighted and appear in the **file name** field lower down the window.

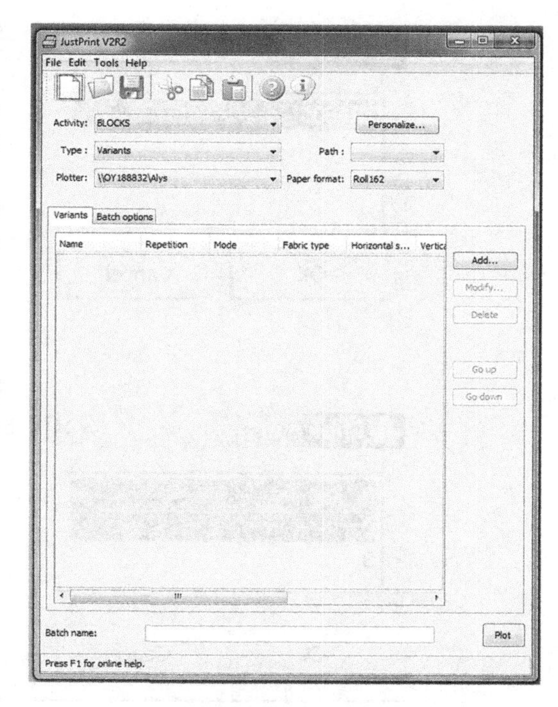

5.3 JustPrint opening window

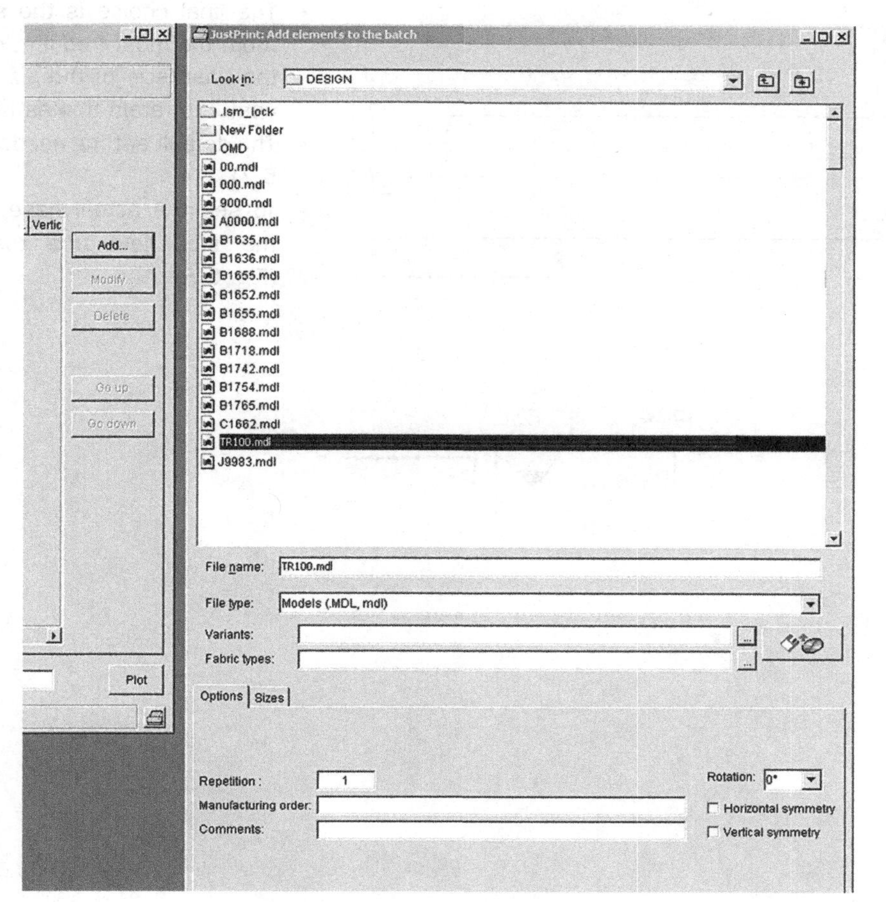

5.4 Add elements to batch window

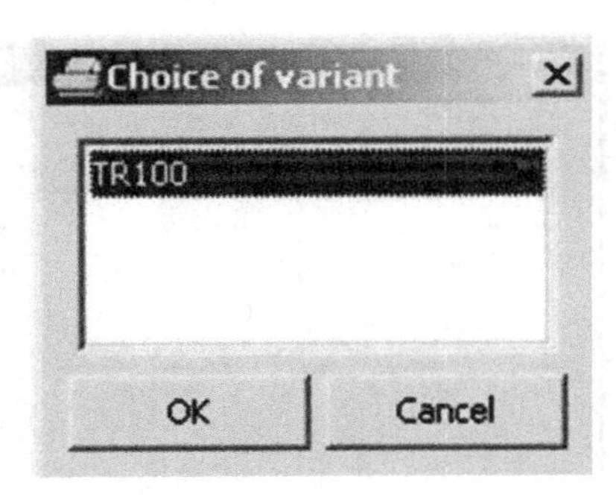

5.5 *Choice of variant window*

5.2.2 *Add variant*

- Use the search button at the end of the **variants** field to access the variants associated with the model. In this example there is just one option, but you may have made more because of a style variation.
- Highlight the option required, click **OK** and it will be entered into the variant field (Fig. 5.5).

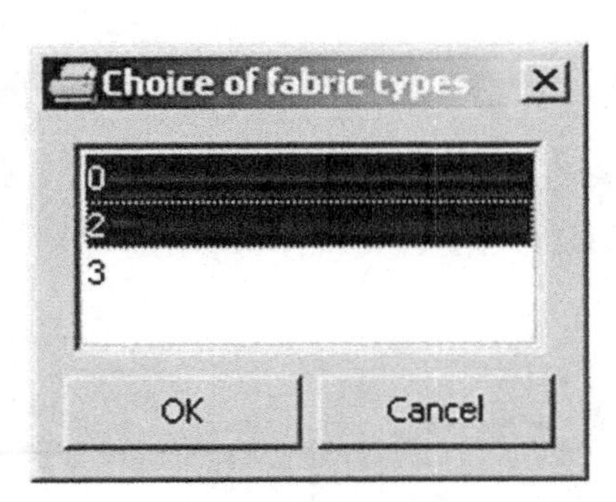

5.6 *Choice of fabric window*

5.2.3 *Add fabric type*

- Now use the search button at the end of the **fabric types** field to access the fabric options. Choose just one option to plot the one fabric type, or multiples to plot all of the pattern pieces.
- Highlight the fabric codes required, and click **OK** (Fig. 5.6).

5.2.4 *Add base size*

- The final choice is the size. In a sample room the most frequent request will be for the base size of the size range, making it useful to retain this as the default setting.
- The default setting needs no selection (Fig. 5.7).
- To see the actual base size displayed in the size field use the search button (Fig. 5.8).

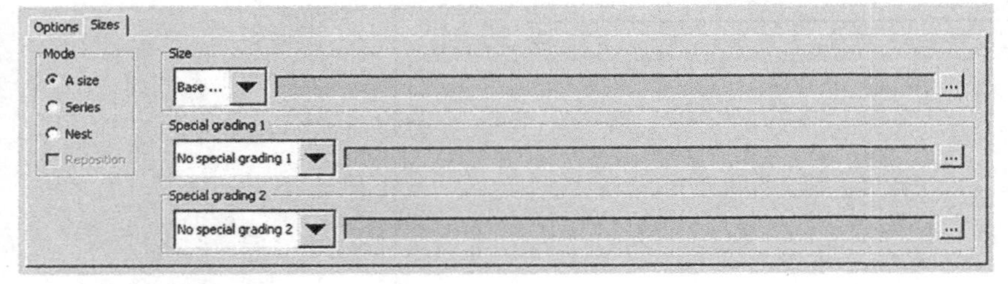

5.7 *Size choice main field*

5.8 *Base size option displaying the size*

5.2.5 *Add other sizes*

At the sample stage it is probable that the model file is ungraded or the grade is still unchecked. If you require a different size to the base size ensure that you have checked the graded pattern, as an ungraded pattern will still plot out with the requested size printed onto the pattern but actually be the base size. To plot a size other than the base size:

- use the drop-down menu arrow in the size field and select **Other** . . . (Fig. 5.9);
- next use the search button to access the available sizes (Fig. 5.10);
- highlight the chosen size and click **OK** (Fig. 5.11).

5.9 Add other size option.

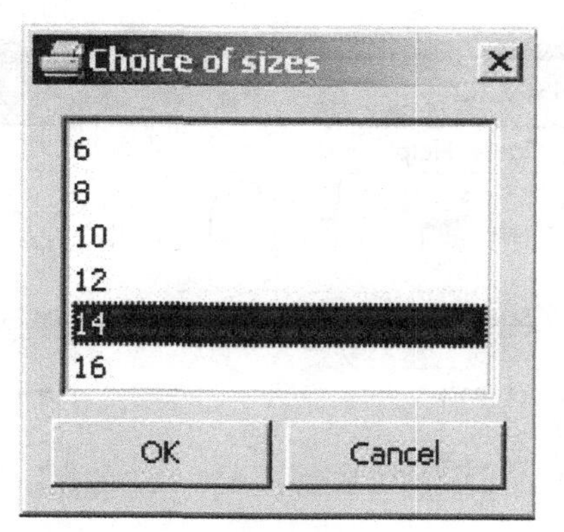

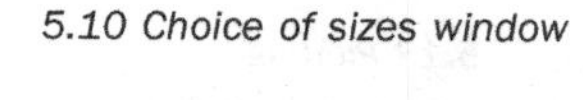

5.10 Choice of sizes window

5.11 Size 14 chosen and displayed

5.2.6 *Send model to plot*

Having completed the order form (Fig. 5.12) click on the **Add to batch** button at the bottom of the window. This will transfer the order to the main JustPrint window.

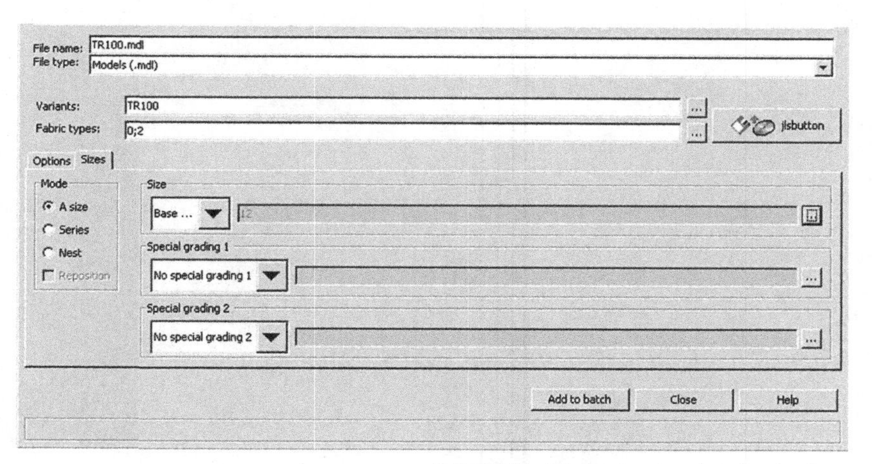

5.12 Add batch to plot completed order form

Finally click on **Plot**, at the bottom right of the window, and the file will be sent to the plotter, highlighting **green** to confirm the action (Fig. 5.13).

Should it turn **red** the action has failed, so go back through the procedure and check that everything has been filled in correctly. A failure to print may be due to the settings, particularly the access paths, or to the variant so check these too.

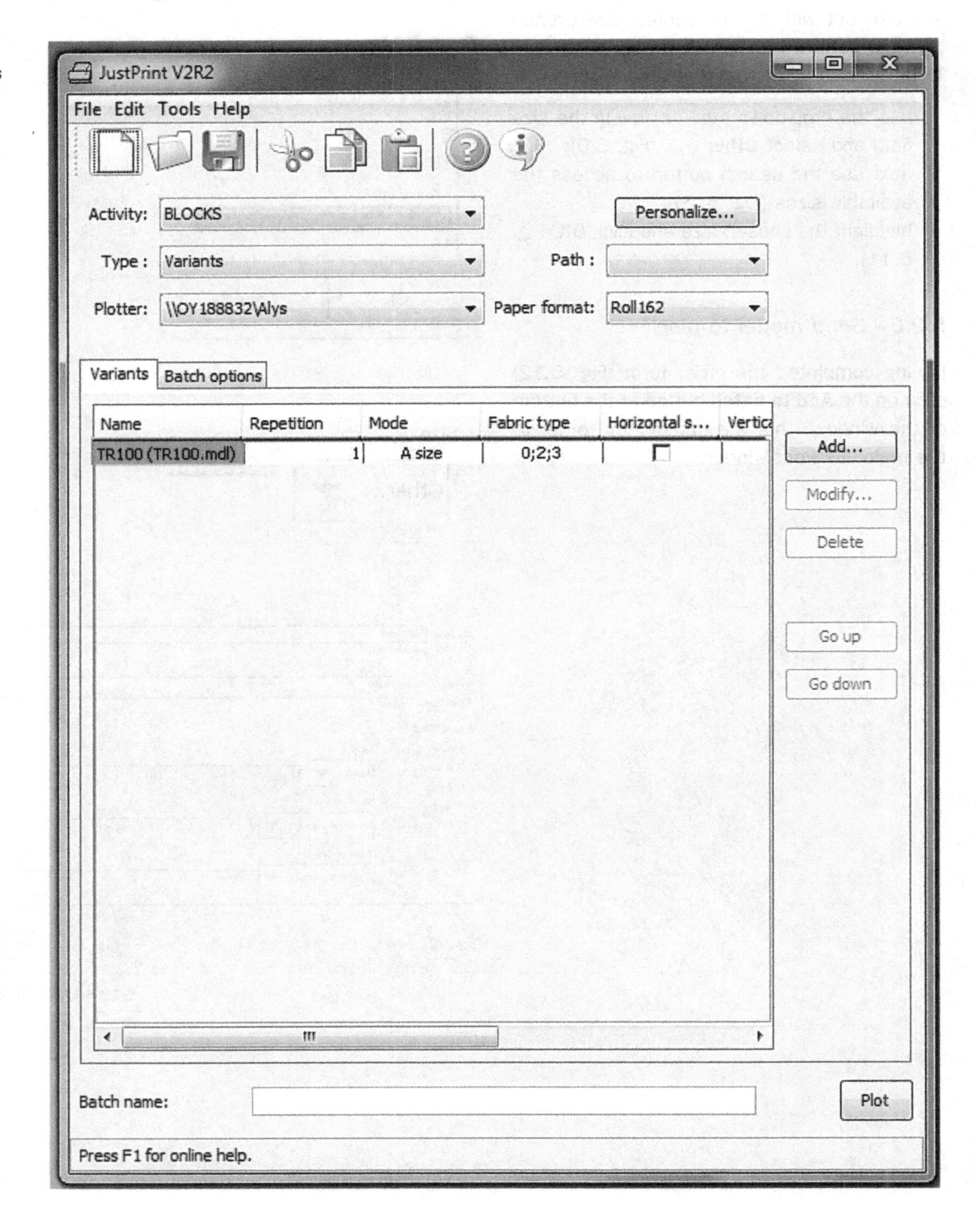

5.13 Plotting successful highlights green

5.3 How to plot single pieces

Use the following procedure.

- Launch JustPrint and on the opening page change the **type** field from **variants** to **pieces** (Fig. 5.14).
- Continue by following the instructions in Section 5.2.1 Add model **file** and Section 5.2.2 Add variant **selection**.

You will see that the **fabric type** field is now replaced by the **piece article** field.

- Use the search button to access the **piece article** list. Highlight the single piece required or hold **Shift** while making a multiple selection, and click **OK** (Fig. 5.15). Please note that **Shift** applies when making a continuous selection. When selecting non-consecutive pieces, use the **Control** key.
- Continue as in Sections 5.2.4 or 5.2.5 to select the size (Fig. 5.16).

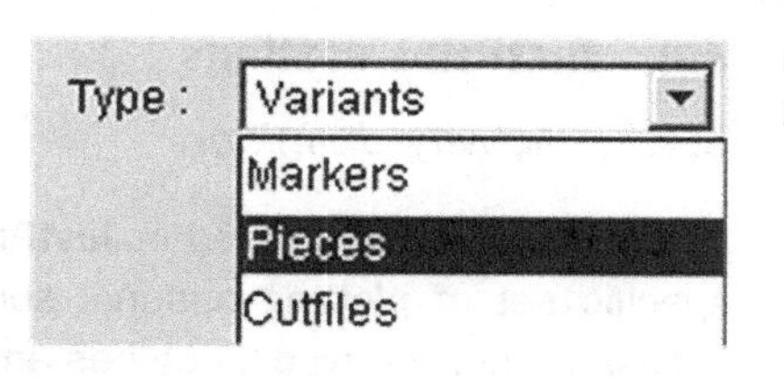

5.14 Pieces selected to plot

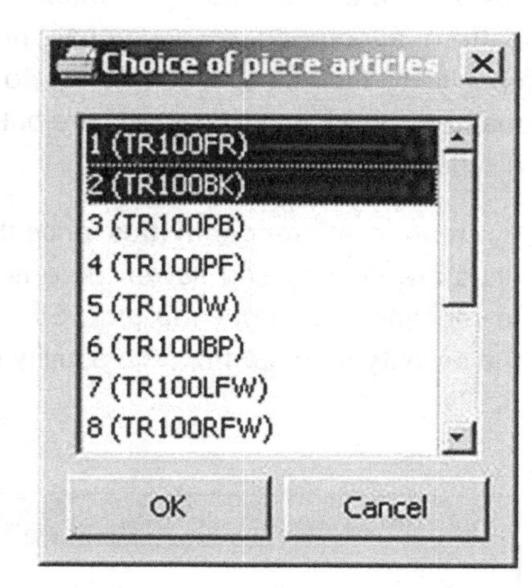

5.15 Choice of piece articles window

5.16 Completed pieces order form

- Finish as in Section 5.2.6, and the selected pieces will be plotted out.

5.4 JustPrint settings

5.4.1 *Activity definition*

An **activity** is a file created in **JustPrint** with a specific set of plotting options, such as the access paths, line types, notches and text and then saved with a name of your preference. For example, you may decide to plot **blocks** with a different set of options to a **model**. Alternatively, different clients may require separate pattern annotation or additional information specific to their files. JustPrint is also used for markers, which is not covered here but requires a different set of options.

By creating different **activities**, once these settings are defined and saved there is no need to redefine each time you plot. Simply select the **activity** required from the activity field (Fig. 5.17).

5.4.2 *How to create an activity*

In this example the activity **Blocks** will be defined, using the Alys plotter and a roll of paper 162 cm wide. Use the procedure outlined below.

- Launch JustPrint.
- Click on **Tools – Activity management** and there are four options (see the right-hand column of Fig. 5.17):
  - **Create** . . . to make a new activity file;
  - **Open** . . . to open an existing activity that can be checked or amended as required;
  - **Delete** . . . as it says, to remove a file;
  - **Duplicate** . . . to copy a file, useful when only a minor change is needed to an existing file, but for a new profile.

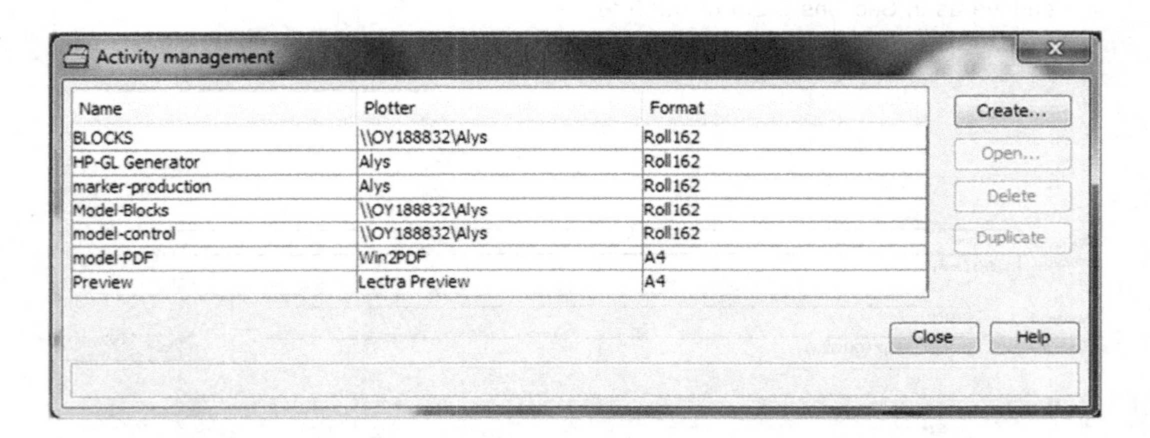

5.17 *Activity management folder*

- Select **Create** and the **Open activity model-control** window will open (Fig. 5.18).

This window contains a number of tabs with various fields in which to define your requirements. Starting at the top there are four initial fields to complete.

- **Activity**: Enter the name of your defined profile. This could be the supplier or customer name, and in this example is **Blocks**.
- **Plotter**: from the drop-down menu select the plotter or printer type required. In this case **Alys**.
- **Paper format**: from the drop-down menu select paper type required, here **Roll 162**.
- **Plots in a file (HP-GL ASTM)**: this small box, (top right) needs to be un-ticked when plotting models on a plotter. However, it could be useful when the file is used to plot on another plotter.

5.4.3 Access paths

Define the **work folders**, in other words the pathways to the required files, and fill the top line first. Use the search button to access and choose from available files, then use the blue arrow to auto fill form. In this example everything to do with Blocks is in the same folder, but for other profiles such as markers the items may be in different folders.

Once the access paths are entered, press **Save path** and a dialogue box will appear in which to name and save the pathway. It is useful to use a path folder name, e.g. design, blocks etc. Enter the name and **OK**. However, note this is not the main save for the page which is at the foot of the window. Points to note:

- leave the path that will be used most often as the default,
- **model server** – leave blank,
- now use the main **save** at the foot of the page.

Note that it is important to make a save on each separate tab.

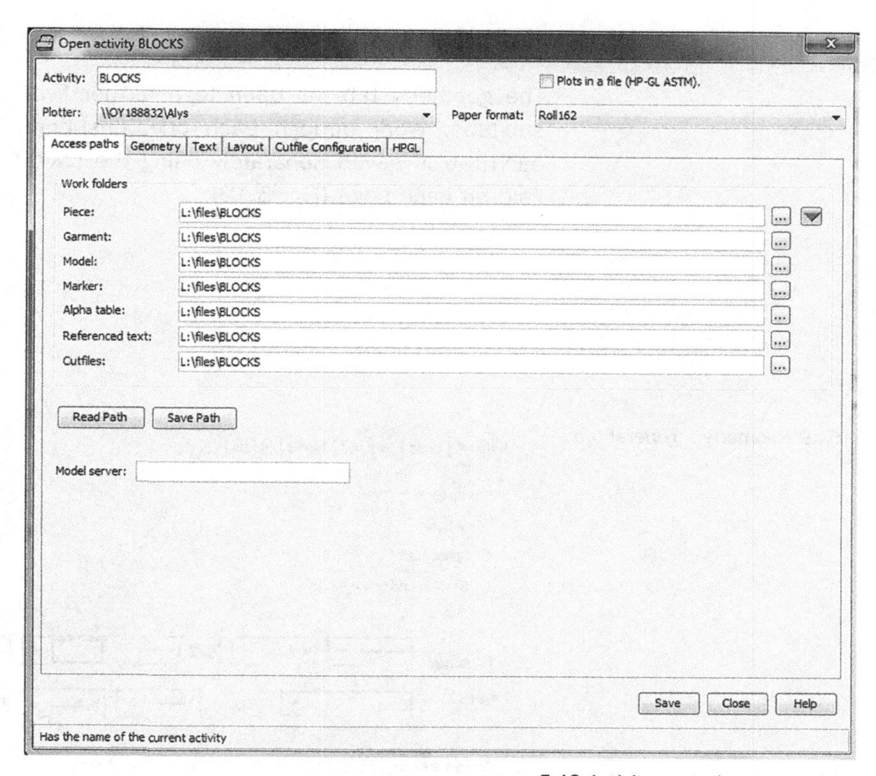

5.18 Activity control opening window and Access paths tab

5.5 Geometry

The geometry tab will open to a further five sub-tabs. Work through each one, ensuring each tab is saved separately using the save field on each page (Fig. 5.19).

5.5.1 *General tab*

Please note:

- **Plot options**: select **Graphic** and **Fusing pieces and fusing blocks**,
- **Border**: leave unticked for models,
- **Fold**: is used for markers but leave as **open**,
- **Configuration DXF**: not required here,
- Now **save**.

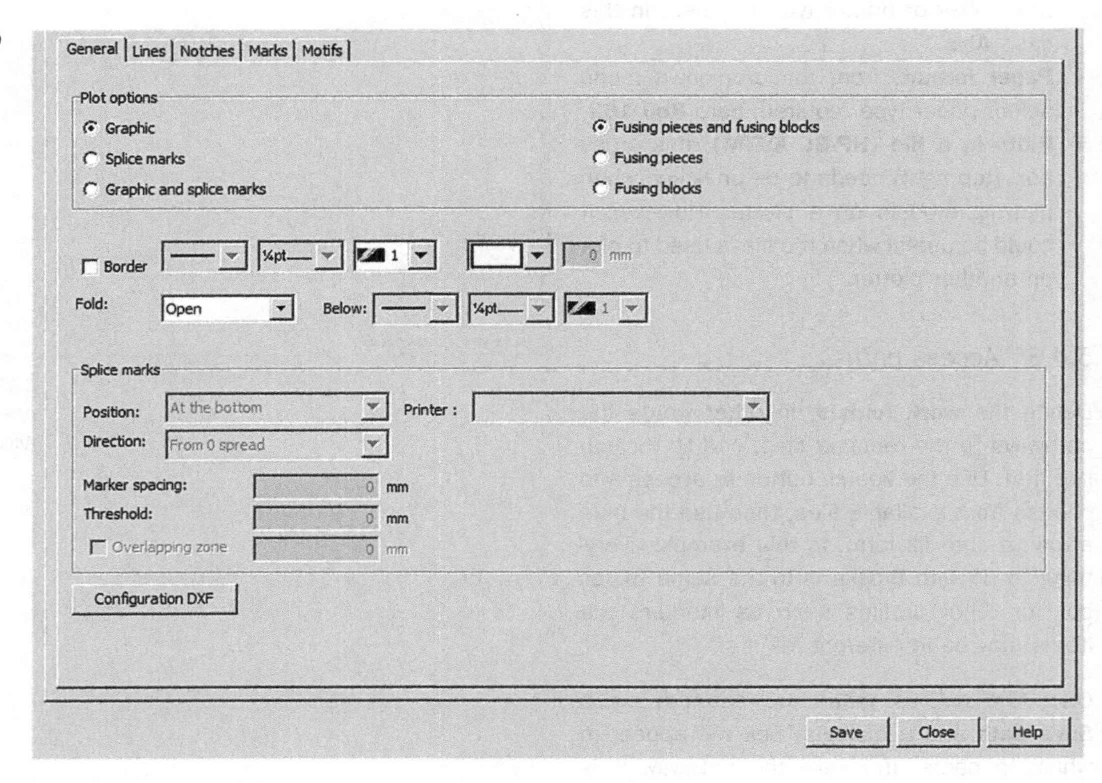

5.19 Geometry – general tab

5.5.2 *Lines tab*

Please refer to Fig. 5.20. This tab has a list of all the lines that it is possible to plot out. Tick and un-tick line selection boxes, and choose the line thickness, solid or dashed as required. The final column allows for colour but normal plotting is in black and white so leave at the default setting.

The outlined below selections work well for normal plotting of models.

- **Contour line** is the outside, cut edge of the pattern. Solid, $1/2$ point line.
- **Grain line** is a solid $1/4$ point line. Select **small** as this will keep it inside the pattern edge. Choose the graphic display for the grain line, arrow, double arrow, line etc.
- **Seam line** is a dashed $1/4$ point line.
- **Internal plot** is a short dashed $1/4$ point line. This is for any internal lines of a pattern other than the seam line.
- Now **save**.

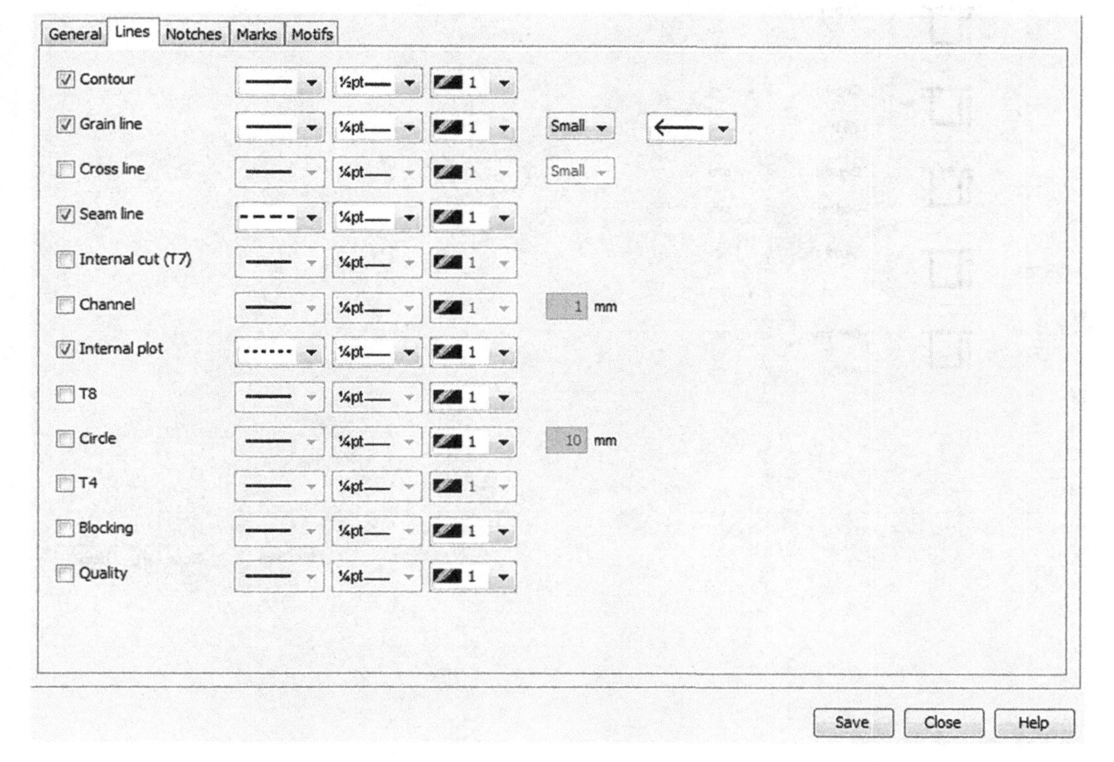

5.20 Geometry – lines tab

5.5.3 Notches tab

Please refer to Fig. 5.21. This is a list of all the possible notch types. Choose the notch type required, and adjust the settings to determine the length and width of the notch. It is possible to enter a value by keyboard, e.g. '0'.

Please note:

- **Notch 21** is selected with a depth of 5 mm, and without any width so will plot out as a line,
- Now **save**.

You can select as many options as you require.

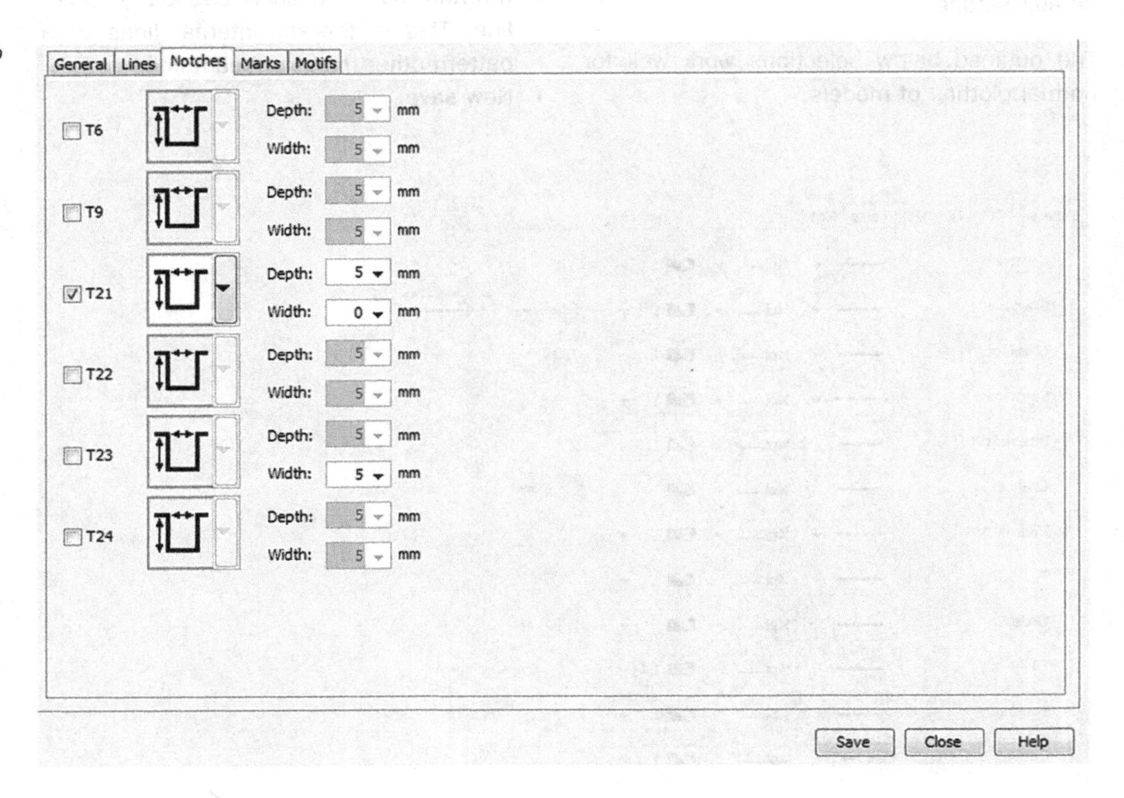

5.21 Geometry – notches tab

5.5.4 *Marks tab*

Please refer to Fig. 5.22. This is a list of possible marks, or drill hole types. Choose the mark type required and then its graphic image plus the diameter required. The final column has a colour option but is not needed for general plotting. Please note:

- **T35** chosen as a cross with a diameter of 5 mm,
- Now **save**.

You can select as many options as you require.

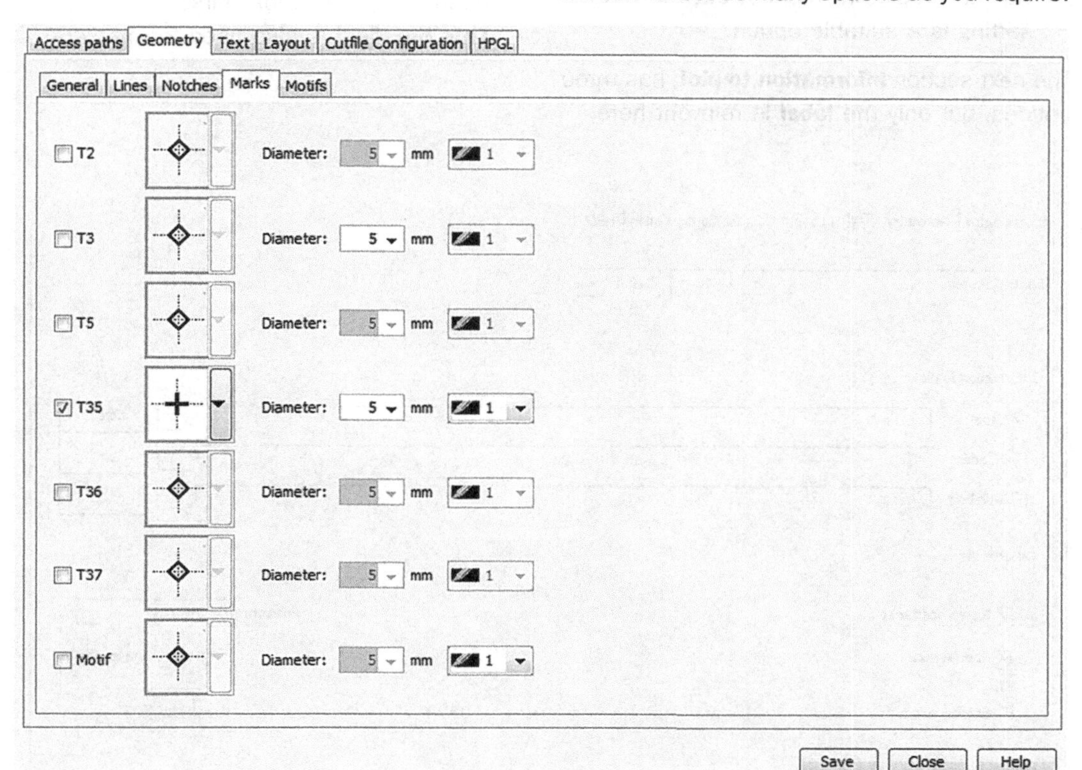

5.22 Geometry – marks tab

5.5.5 *Motifs tab*

Please refer to Fig. 5.23. Motifs are used for locating repeat prints or placement prints etc. in Diamino and are not needed when plotting models.

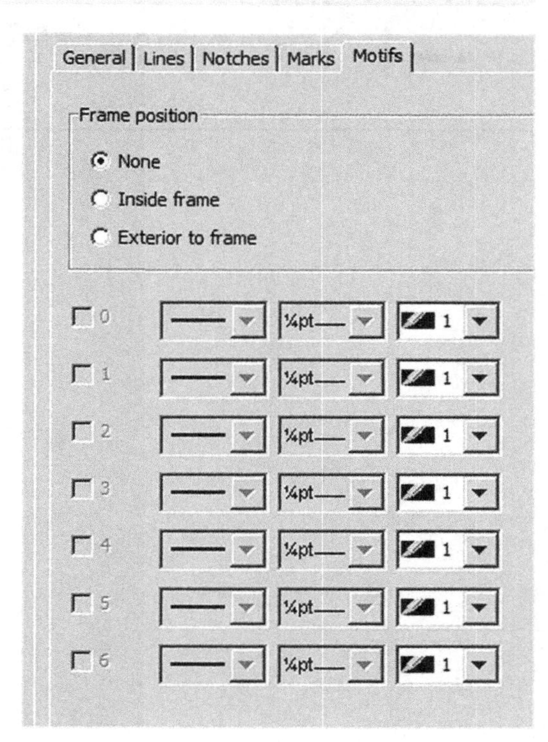

5.23 Geometry – motifs tab

5.6 Text

Please refer to Fig. 5.24. The text tab is where both the 'what' and the 'how' text are added to the pattern when plotted out:

- **Font:** Select font and point size as required via the drop-down menus, but the default setting is a suitable option.

The next section **information to plot**, has three options, but only the **label** is relevant here.

- **Label** is the internal label or text written onto the pattern piece. First the label is created and saved the as a .lab2 file extension for JustPrint. The label is then selected from the Textlib file and displayed in the label field.
- **Header** field used for markers.
- **Title box** field is also used for markers.

5.24 Text tab

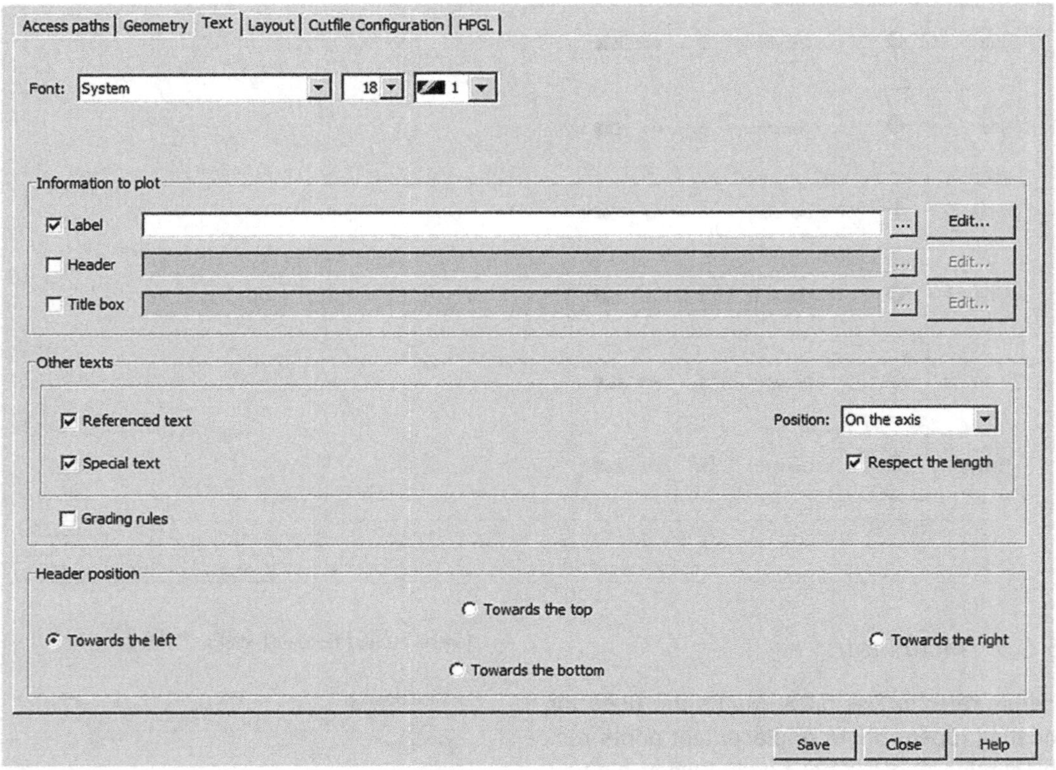

5.6.1 *To create a **label***

Use the **Edit** . . . button at the end of the **label** field to open to a new window in which to select information to be printed onto the pattern. Note that this information has already been attached to the model by the use of the text boxes, or file names and this field allows the importation of the text (Fig. 5.25).

Use the following procedure.

- Use the menu on the right-hand side to choose between **general**, **model**, **piece** or **marker** information.
- Select the required information that you want to print onto the pattern from the list displayed below the chosen option.

The points outlined below should be noted.

- The example shows options chosen one at a time and then the blue arrow at the bottom right of the window used to transfer the text into the main label field. This results in the information laid out as a block.
- It is possible to select everything to be on the same line.
- It is also possible to customise the font, size of font and position on label. Use the on-line help index and experiment with the toolbar at the top to make a customised display.
- Save the label to a label file library with a .lab2 file extension for JustPrint.
- Close the window.
- Now use the **Search button** at the end of the **label field** to access the label file and select. It will be displayed in the label field.

5.6.2 *Other texts*

'Other texts' refers to the **referenced text** and **special axis** created in Modaris using F4 – Axis. Please note the following:

- tick the boxes as required,
- **Position** refers to the placing of the text on, above or below the axis,
- **Respect the length** – means that text will stay inside the piece, so keep ticked,
- **Grading rules** have not been covered here, and are not required,
- **Header position** is for markers,
- Now **save**.

5.25 Creating a text label

5.7 Layout

This section is about the gaps between pattern pieces and the paper edge, gaps between the pieces themselves and an option to scale the pattern when sending to an A4 printer (Fig. 5.26).

Please note the points outlined below.

- **Margins** not needed for general plotting so leave as **0**.
- **Spacing in mm** is used to define the space between pieces as they are plotted out in both the ***x*** and the ***y*** direction. A small gap is useful but you can adjust to your needs. In this example 10 mm has been allowed between pieces.
- **Plot position** is used to determine where to place the pieces on the paper. Select **Bottom alignment**.
- **Reduce in page** should be left unticked for normal plotting. It is used to scale a pattern to fit onto one page when sending to A4 or A3 printer.
- Now **save**.

5.8 Cutfile configuration and HPGL

These two tabs are for more complex actions and should be left in the default setting for general model plotting.

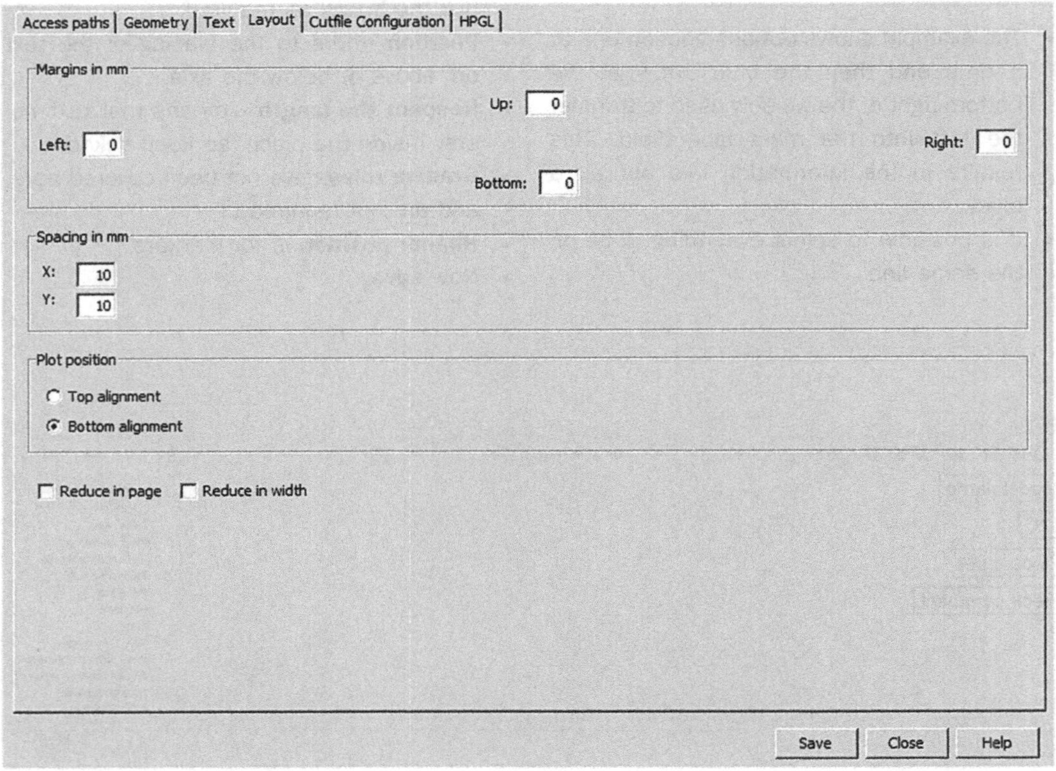

5.26 Layout tab

Chapter 6

A guide to the toolbox in Lectra Modaris pattern cutting software: F1–F3

Abstract: The purpose of this chapter is to introduce the Modaris user to the specific pattern cutting toolbox displayed on the right-hand side of the screen, also known as the trade functions. Each menu is explained in turn, working through the functions from the top of the list down. Links are made with the tool bar menus and the keyboard. Only those most commonly used for pattern cutting are explained here, i.e. the commands that you need to know, what they do and where you would use them.

Key words: dialogue box, associated parameters, points, lines, notches, seam allowances, menus F1, F2, F3, Lectra Modaris.

6.1 Introduction

This chapter focuses on the Lectra Modaris V6 function menus, also known as the trade functions, that are displayed on the right-hand side of the screen. The commands are explained by taking each menu in turn and working from the top of the list down. Because some functions duplicate others, or are for complex procedures, only the most commonly used functions for general pattern cutting are explained here, i.e. the commands that you need to know, what they do and where you would use them. Consequently, some functions are omitted. There are some commands that will be familiar for instance the use of the left and right mouse buttons to select an item and then finish the action or Ctrl + Z/z to undo the last command. Other functions require a series of actions, such as selecting the item or items first and then completing the action.

To use a function, first select the specific function menu, **F1**, **F2** etc. and then click on the relevant **command field**, both of which are written in bold type. Then carry out the procedure that immediately follows.

References are also made to Chapter 8 where further options are available within a function by the use of the upper tool bar menus. Keyboard use is also an integral part of the system, and this is explained with the specific function.

6.1.1 *Dialogue box*

The **dialogue box** is an important addition to many functions allowing an exactly measured action. Figure 6.1 shows a sample of the various formats, some requiring one measurement, others a number of measurements.

Diameter will be displayed when creating a circle in **F2**.

Beginning and **end** refer to the line, moving in a clockwise direction when adding seam allowances in **F4 – Seam**.

Width and **height** refer to the sides of a rectangle created in **F2**.

Diameter 80 mm Lock

Beginning 10.00 mm Lock
End 10.00 mm Lock

Width 200.00 mm Lock
Height 100.00 mm Lock
Rotation Decimal degree Lock

dx 111.12 mm Lock
dy 22.59 mm Lock
dl 113.39 mm Lock
Rotation 90.00 Decimal degree Lock

6.1 Dialogue boxes

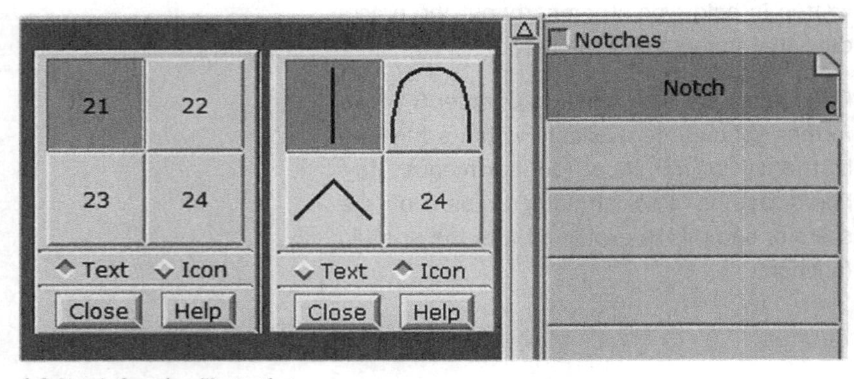

6.2 Notch function illustrating associated parameters

The **dx, dy** and **dl** fields refer to the direction of the action when moving or adding points and lines:

- **dx** moves on the *x*-axis, i.e. along the horizontal. Enter a minus (–) to move to the left,
- **dy** moves on the *y*-axis, i.e. on the vertical. Use a minus (–) to move down,
- **dl** is the straight line measurement directly between two points.

Rotation is in relation to the starting point and fills in automatically.

How to use the different formats will be covered with the specific command. In general:

- read the **dialogue box** that opens and respond accordingly,
- use the **arrow keys** to move between fields,
- use **enter** to complete the action.

If the **dialogue box** is in the way, use the blue bar across the top of the box to **click and drag** to a more convenient location. Alternatively, an action can be completed freehand by simply ignoring the **dialogue box** and clicking at the required position.

6.1.2 *Associated parameters*

Associated parameters are the sub-menus of a specific command showing the extra variations available within that function. Functions with this facility are indicated by, and accessed through the corner tab of the specific function.

Figure 6.2 displays the associated parameters of the **notch** tool. It shows that notches 21, 22, 23 and 24 are available, and by selecting **text** or **icon** the menu can be displayed either way. Select the **notch** of choice and **close**. **Notch 21** is common for general pattern making, but can vary with product type. An alternative, but less efficient way to access **associated parameters** is through the drop-down menu **parameters**, see Section 8.11.4

6.1.3 *Text or icons?*

Ctrl + **S** will change the right-hand command column from **text** to **icons**, and back again. When selecting the icon display, other sub-menus will also change to display the icon option (see Fig. 6.3).

Configure-icon/text on the upper tool bar is an alternative method but the keyboard command can be activated unintentionally thus making it difficult to identify the cause when your screen display changes.

Use **Ctrl** + **S** to change back again.

6.1.4 *Shift and Control*

The **Shift** and **Control** keys vary some commands. Functions with this option will be identified along with how and why they are used.

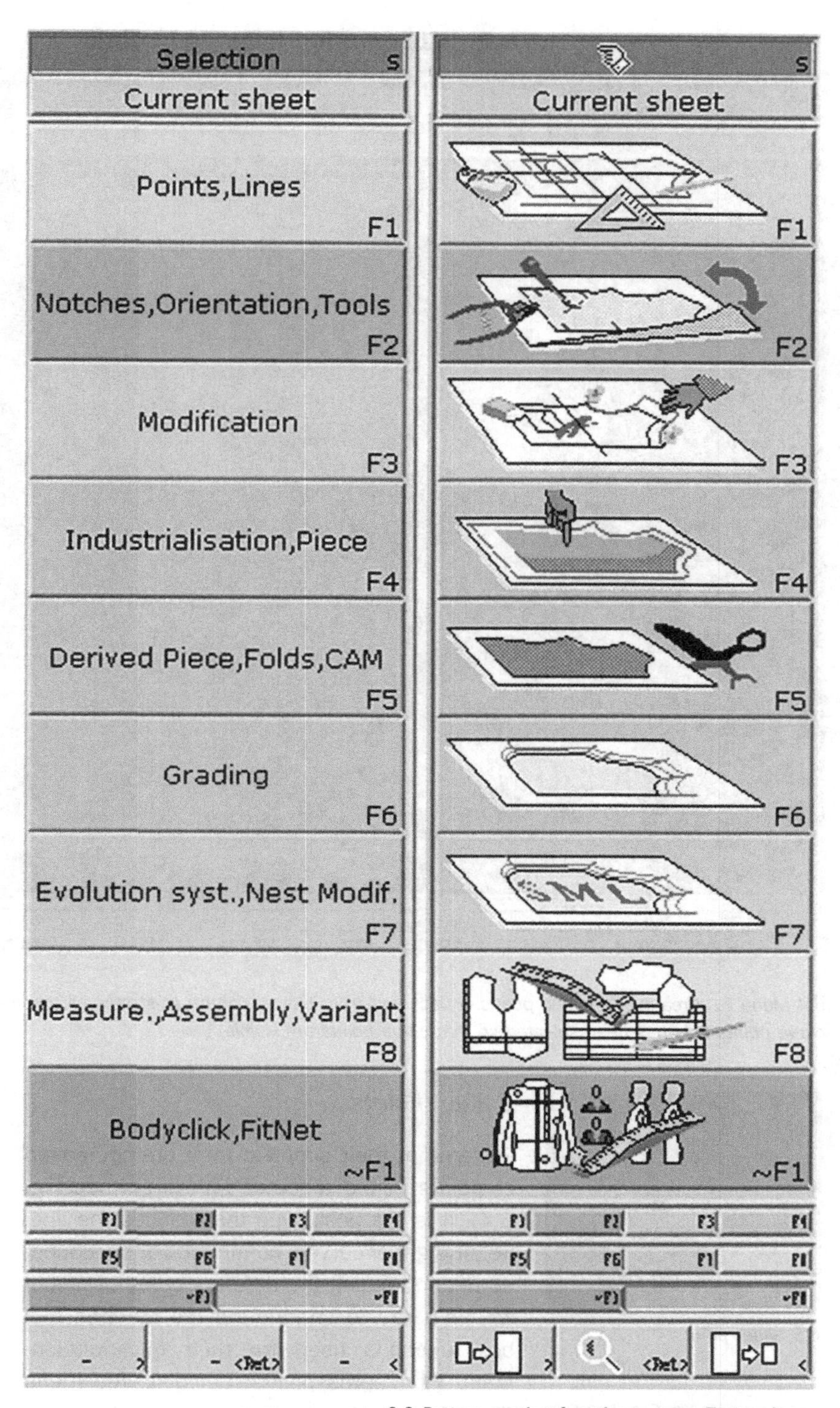

6.3 Pattern cutting functions menu. Text or icons

6.2 F1: creating lines and points

6.4 Menu F1. Creating lines and points. A back and front blouse pattern illustrates all the lines and points used in general pattern making. The curve points field has been selected so that curve points are visible

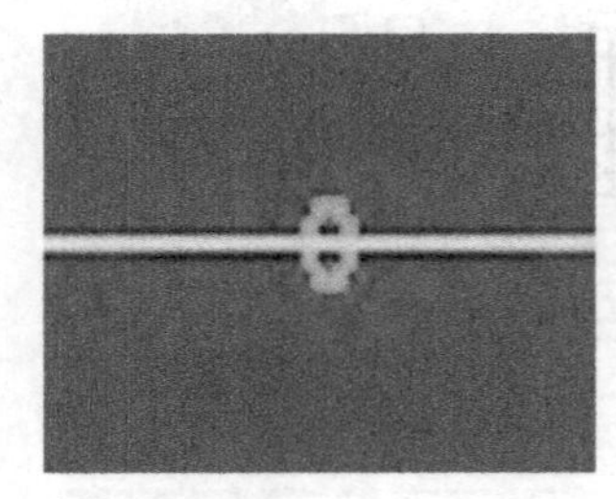

6.5 Slider point

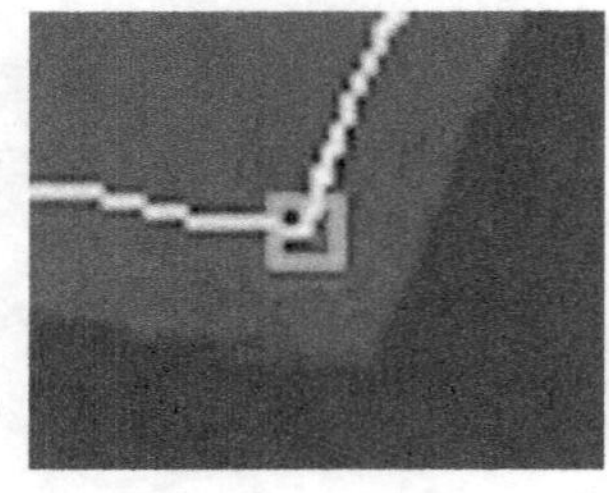

6.6 Corner point

6.2.1 Points

Patterns at their simplest form are composed of points joined together with lines (see Fig. 6.4). It is the point type that defines the line, i.e. straight or curved, continuous or sectioned, so understanding the different types of points will help you to understand the Modaris tool box functions used for their manipulation. Menu **F1** creates points, but also see menu **F3** on how to change from one type of point to another.

6.2.2 Creating and adding points

Slider point (Fig. 6.5) is shown as a loop and can be moved along a line without distorting or altering the shape of the line. It will also retain its relative position when the end points of the line are moved.

F1 – Slider points are added by simply clicking at the required position.

Use **F3 – Reshape** to slide the point along the line.

Use **F3 – Insert point** to transform the slider point into a characteristic point.

Corner or End point (Fig. 6.6) (key 2 on the digitising handset) is shown as a square, and defines the ends of lines. Corner points are created only when digitising, or developed from an existing characteristic point using **F3 – Section**.

Use **F3 – Merge** to change a corner point into a characteristic point.

They are mostly used at a corner position but can be used anywhere to define a specific length of seam. See Fig. 7.5 where it is used along a straight line to enable a stepped seam allowance.

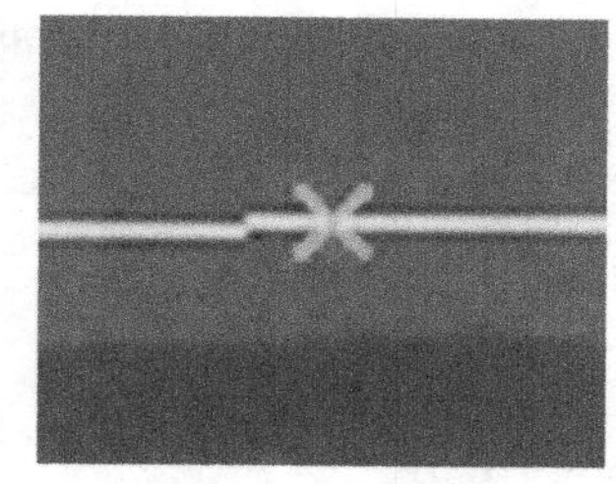
6.7 Characteristic point

Characteristic point (Fig. 6.7) (key 1 on the digitising handset) is shown as a white or blue cross on screen and respects the sharpness of a line. Blue just indicates that the point has been graded or associated to a graded point, but always check a grade, as it may become distorted during pattern manipulation.

Use **F3 – Section** to change a characteristic point to an end point.

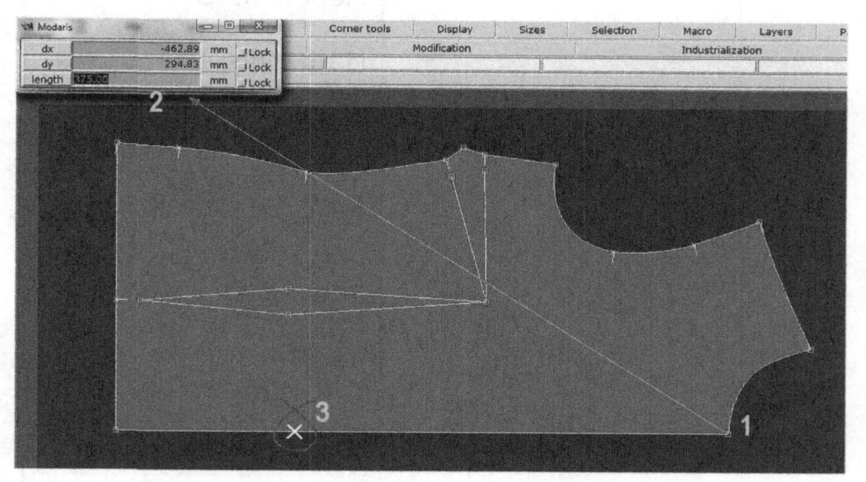

6.8 Add a point using the dialogue box

F1 – Add point Please refer to Fig. 6.8. To add a point to a line, first click on an existing point because the function needs something to copy. In the example in Fig. 6.8 it is the CF neck point (1). Once selected, both an arrow and the **dialogue box**, (2) top left of screen, will appear. There are now two options.

- Enter a value into the dialogue box equal to the distance from the copy point to the required new position, in this case 370 mm. Note that this value goes in the lowest field. Use the **arrow keys** or drag the cursor to the **length** field, click and enter the value. Then click anywhere in the general direction of the new point position, but on the required line, and the new point (3) will be placed exactly at the distance specified.
- Alternatively, simply click on the line where you want to add the new point (2), using the on screen ruler and a freehand action (Fig. 6.9).

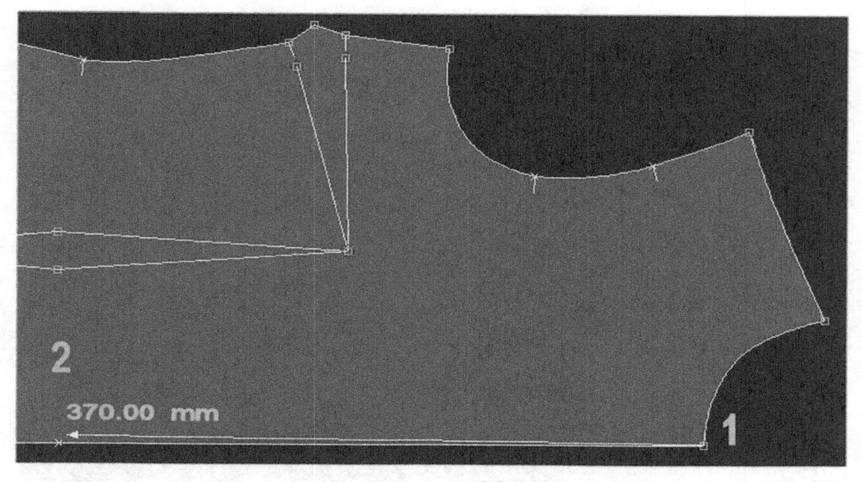

6.9 Add a point using a freehand action

F1 – Add point + Shift is used to create a **curve point**. Generally there is no need to specify the exact position for a curve point; it's a simple click while holding down **Shift**, and insert freehand.

Curve points (Fig. 6.10) (key C on the digitising handset) are shown as a red cross on screen, and are not visible until the **curve points** tab found along the base of the screen is selected. Use the fewest possible **curve points** and in uneven amounts, to obtain a good curved line. **Curve points** are not graded and should not be used for notches.

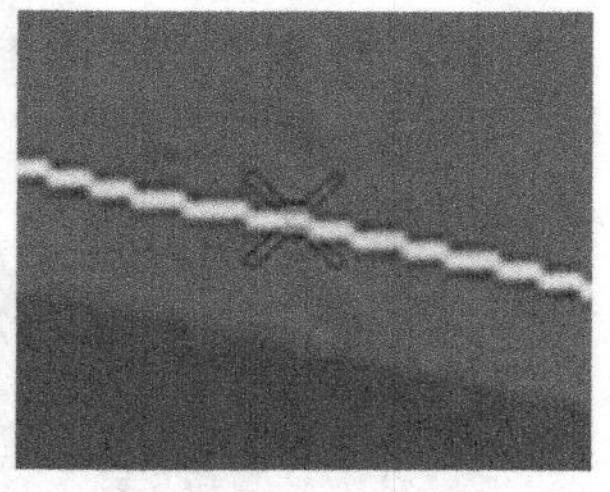
6.10 Curve point

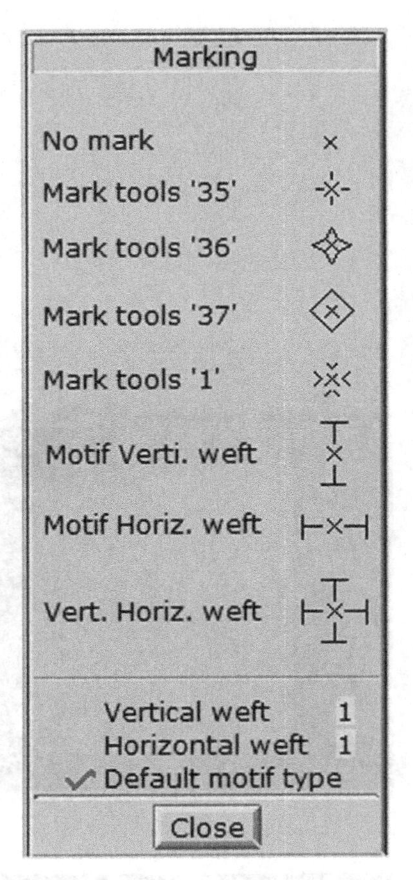

6.11 Relative point graphic options

6.2.3 Relative points

Relative points (key 3 on the digitising handset) are used for marking drill hole positions at dart ends, button or pocket placements etc. (Fig. 6.11). They are placed within the pattern perimeter and therefore placed *relative* to other points. There are a number of display options for these points illustrated in Fig. 6.4 down the centre front button marking positions. The choice of graphic display (associated parameters) is accessed via the corner tab of the menu, but the way to create a relative point is the same for all.

Select **F1 – Relative point**, then click on an existing point and both an arrow and the **dialogue box** (top left of screen) will appear. There are now two options.

- Use the **arrow keys** to select and then enter dx and dy values into the dialogue box and press **Enter**. Both positive and negative values can be used according to the position of the reference point.
- Select the position freehand while using the on-screen ruler if required. Click again to complete the action (Fig. 6.12).

The **no mark** option will actually give a simple blue cross on screen, but will not be plotted out. This is helpful for marking the bust point or other key reference points for on-screen pattern use only.

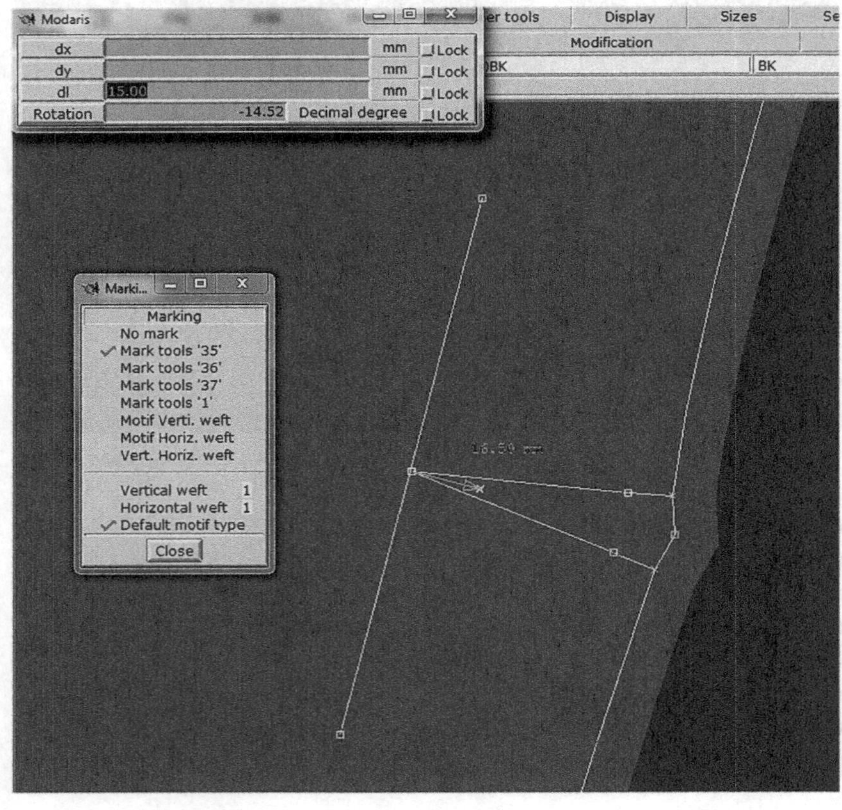

6.12 Adding a relative point

6.2.4 *Aligning points*

F1 – Ali2pts This function allows you to reposition a point to be exactly on the same *x*-axis or *y*-axis relative to another point, in other words, straighten up the line to the vertical or horizontal axis (see Fig. 6.13).

- Click on the point that is in the correct position (1). The arrow indicating the direction of the movement is displayed after the first click.
- If the action is ambiguous, and the movement could be on the *x*-axis (A) or the *y*-axis (B), use the **Space bar** to swap between the options.
- Click on the point to be moved (2).

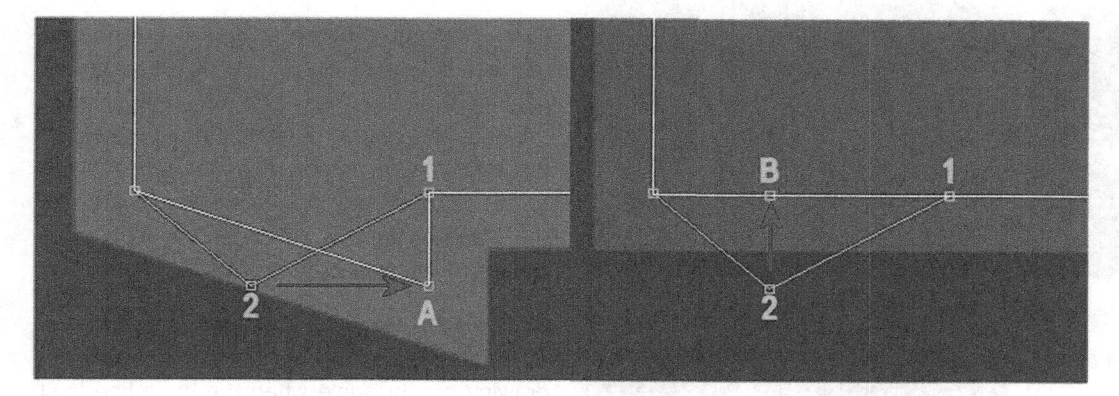

6.13 Align 2 points

F1 – Ali3pts This allows you to realign a point on any axis, unlike Ali2pts that only works on the *x*-axis or *y*-axis (Fig. 6.14).

- Click on the first correctly placed point (1).
- Next, click on the second correctly placed point (2).
- Finally, click on the point to be moved (3). This point will then be realigned to be on the same axis as the other two points.

Note that the point to be moved does not have to be between the first two selected points, and can be on the extension of the first two points.

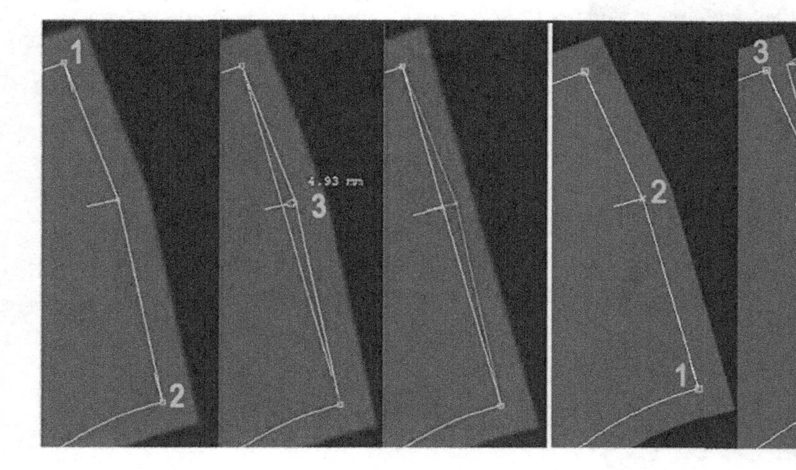

6.14 Align 3 points

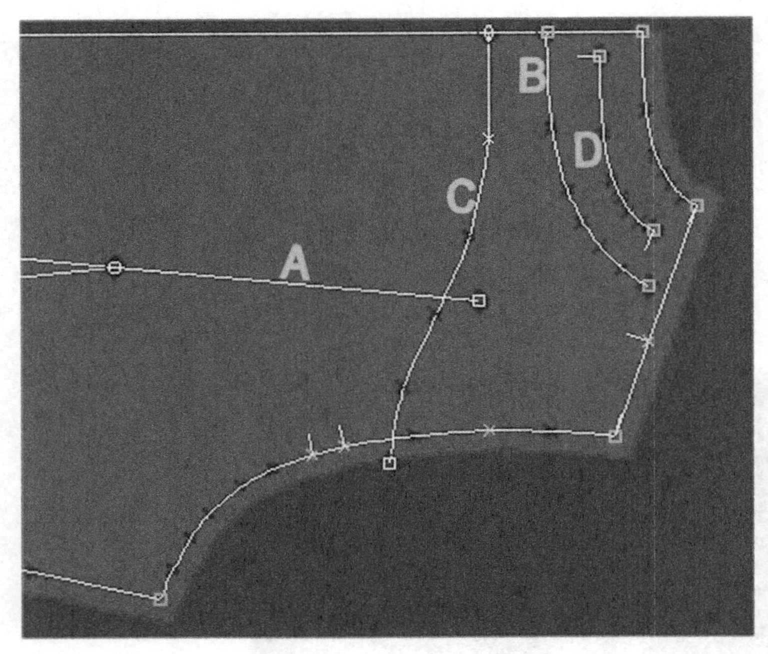

6.15 Creating lines

6.2.5 *Creating and copying lines*

The line functions in this menu create new internal lines on an existing piece, or can be used on a **new sheet** to create a new pattern piece.

Figure 6.15 shows created lines, A = straight line, B = a line parallel to the neckline, C = Bezier line, D = a duplicated line, exactly the same as the neckline.

F1 – Straight creates a new line wherever needed, see Fig. 6.15, line A. Click at the beginning of the line where required, then at the end. This can be done as a freehand action guided by the on-screen measurement line and arrow. Or, after clicking on the first point, use the **dialogue box** to enter precise **dx** and **dy** values for the placement of the end point.

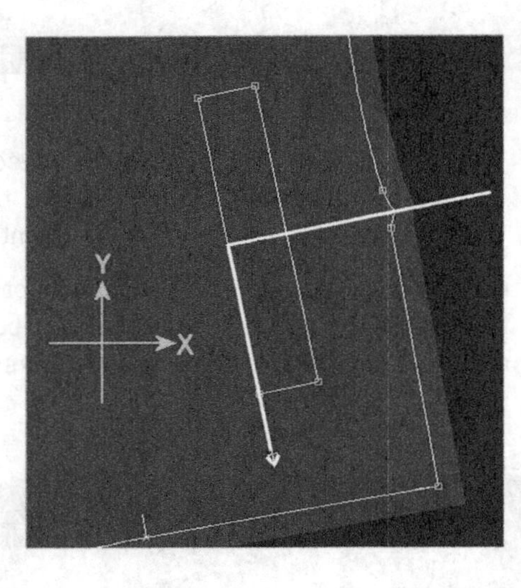

6.16 Create a line at a right angle to another line

Shift + **F1 – Straight** will create the new line at a right angle to the existing line, useful for drawing a right angle when the base line isn't on the *x*- or *y*-axis. Hold **Shift** and click on the first point of the line and drag the cursor in the direction required and the lines will appear on screen. Click again to complete. This function tends to place the new line slightly away from the chosen point, so use **F1 – Parallel** to fine-tune the position and then delete the construction line (see Fig. 6.16).

Ctrl + **F1 – Straight** When creating a straight line this function will keep it on the true horizontal or vertical. It will also keep a perfect 45° angle too. Pull the line to any angle other than 90° and it will lock onto the true bias line. Note that the direction of the angle is defined by the position of the cursor. After the first click, hold **Control** and move the cursor in different directions to obtain the right angle. Figure 6.17 shows the three possibilities.

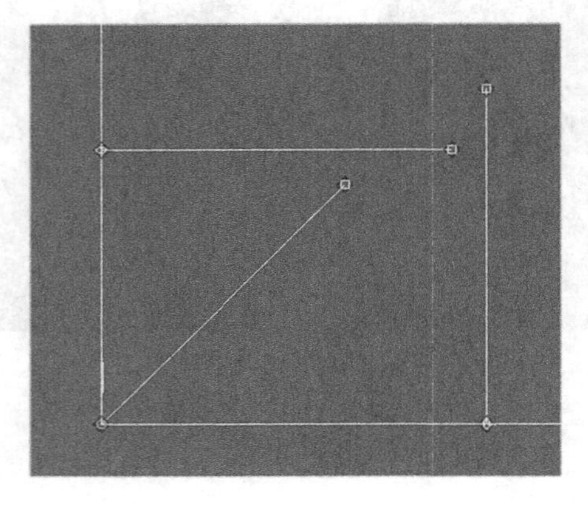

6.17 Using Shift to create lines on x-axis, y-axis or bias line

F1 – Parallel This function creates a new line parallel to the first selected line, see Fig. 6.15 line B parallel to the neckline. Note that if a curved line is copied, the new line will alter and possibly distort to remain parallel. If you require the line to be exactly the same, use **F1 – Duplicate**.

F1 – Bezier This function is used to create an internal curved line, useful for creating design lines on screen, see Fig. 6.15 line C. It is similar to digitising, in that multiple points are created to form the line. Left click at the beginning of the line, and then continue with the left mouse button to create **characteristic** points or hold the **Shift** key for **curve points** as required, forming the curve. Right click to complete the function.

Use **F3 – Reshape** to reposition individual points.

Use Bezier curve **handles** (in the **display** menu) to fine-tune the curve without moving points, see Chapter 8, Section 8.6.8.

F1 – Duplicate is used to replicate a line, selection of lines, points or axis exactly and place them in the new required position, see Fig. 6.15 line D. This action may be made on the existing sheet or can be used to copy to another sheet. Click on the item to be copied and then click again at the new position. Multiple lines can be selected by holding **Shift**, while using the right mouse button. Release **Shift** before completing the action. This is a useful function to duplicate a special axis containing text, such as 'fuse' or 'lining' that is needed on more than one piece. Create and edit the **special axis** first then duplicate as required.

6.2.6 Dividing a line

F1 – Sequence division Use this function to divide a line into a number of equal sections. Click on the two ends of the line to be divided and a **dialogue box** will open. Enter the required number of divisions and click to finish. The points created are slider points. To attach them to the line permanently use **F3 – Insert point**. This is a handy way to space and mark button placements (Fig. 6.18). Also use for finding the centre of a line, dividing a section for pleats or for gathering.

6.18 Sequence division

6.2.7 Digitising

F1 – Digit Select this function to enable the digitisation function. When **Digit** is selected it does not highlight as other function buttons, but instead an on-screen message will read **2 point for horizontal axis**. Digitising can then commence. When digitising is completed (**FF** on the digitising handset) the **F1 – Recover digit** will be highlighted, see Chapter 4 for the full process.

F1 – Recover digit See above.

6.3 F2: notches, orientation and geometric shapes

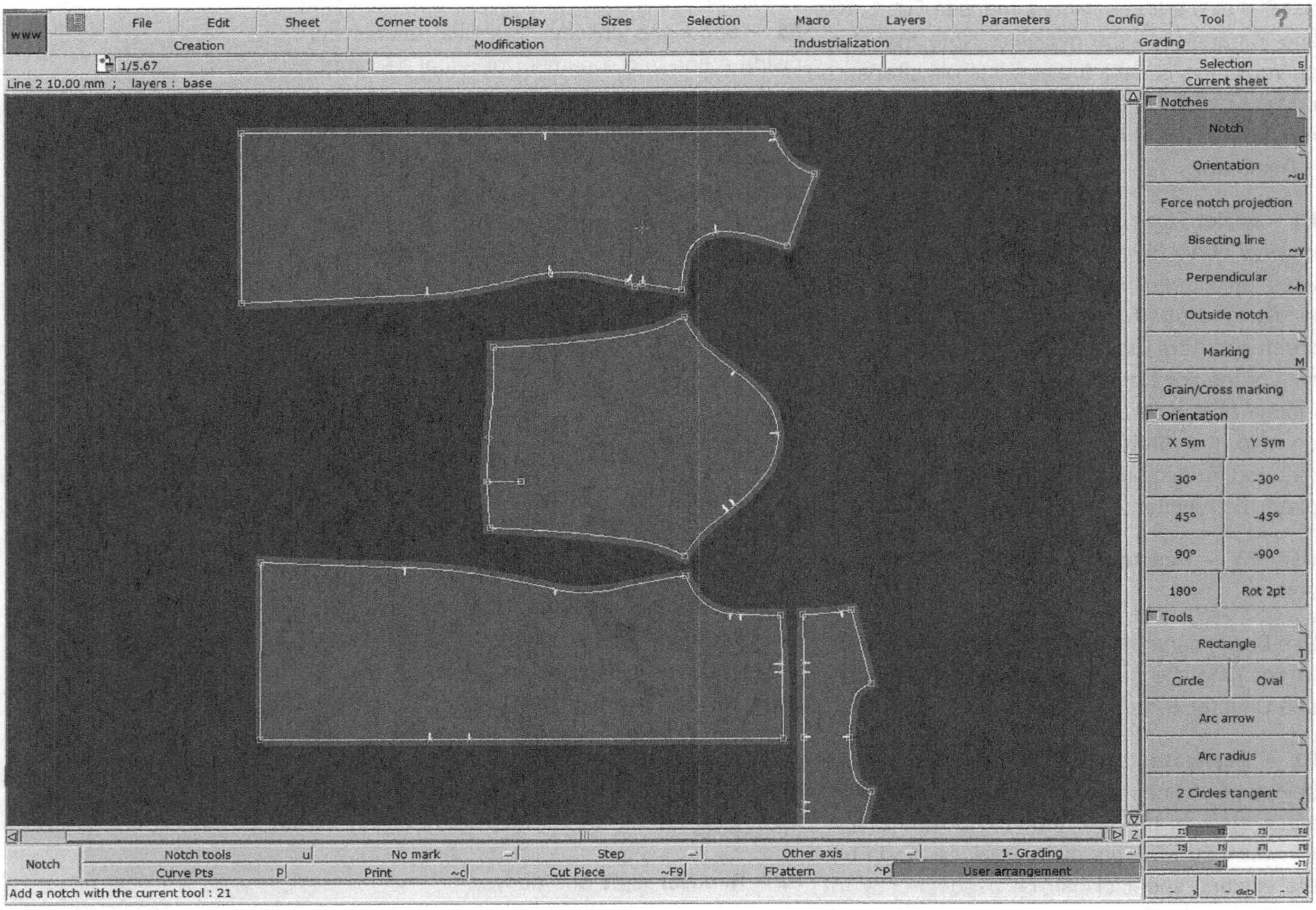

6.19 *Menu F2. Notches, orientation and geometric shapes. A dress with a sleeve illustrates notch groupings and placements*

6.3.1 A note on notches

Ultimately, notches are cut into the fabric and used by the machinist to understand the construction of the garment and sew it together correctly. This is their purpose. Notches are used:

- to identify seam allowance widths;
- as balance marks for matching seams that will be sewn together;
- to distinguish special sewing sections such as darts, ease or gathers.

There is no one set of rules for the use of notches but there are some conventions, and it is the Pattern Cutter's job to place notches in convenient and unambiguous positions.

- Use a single notch on the CF seamline and increase towards the CB.
- On a sleeve head use a single notch at the front and a double notch at the back.
- Space double notches no closer than 1.2 cm. Any closer may not be clear when cut in the actual cloth, and will also weaken the seam.
- By varying the space between double notches they can be used to identify different sections.
- Only use one notch at a corner, placed in the direction of the first line of sewing. A notch at each side of a corner will weaken the fabric and it may tear.
- Do not place notches in the middle of a line except to identify the CB, CF or the middle of mirrored pieces.

F2 – Notch adds a notch to an existing point or directly onto a line, in which case it will be a **slider**. Sliders are useful when matching notches on different pieces as they can be moved without altering the shape of the line, and then attached to the line by using **F3 – Insert point**. See Section 6.1.2 **associated parameters** for the different notch types.

F2 – Orientation of a notch (not to be confused with the piece orientation functions below) is used to realign the notch. Click on the notch and pull in the required direction (Fig. 6.20).

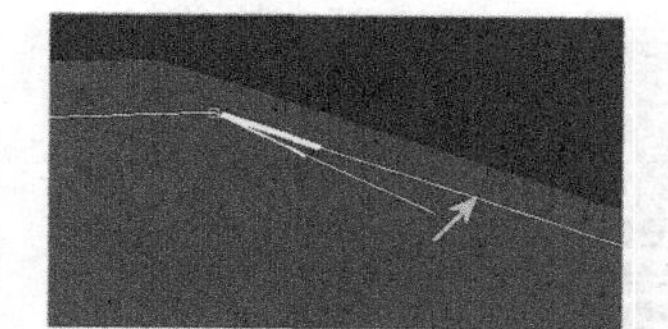

6.20 F2 notch orientation

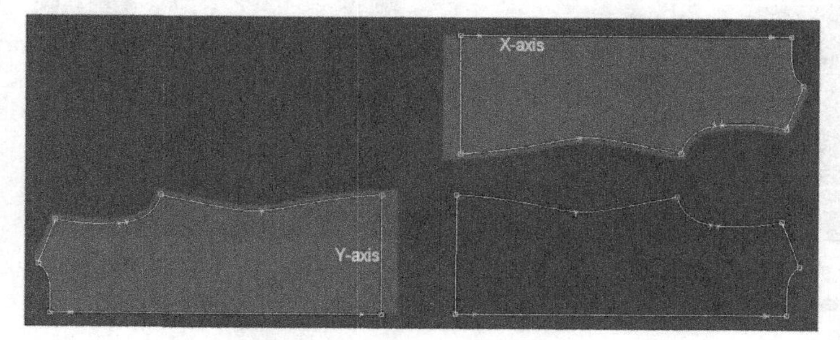

6.21 Flip a piece using X + Y Sym

6.3.2 Orientation of a piece

This section of the menu allows the rotation of a piece while on its own sheet. When making separate left and right pattern pieces, for example a jacket front with different pockets on each side, complete one side, make a copy and flip on the *x*-axis to create the left side before completing the pattern piece variations.

F2 – X Sym will flip the selected sheet on the *x*-axis.

F2 – Y Sym will flip the selected sheet on the *y*-axis (see Fig. 6.21).

F2 – 30°/45°/90°/180°/–30°/–45°/–90°. These functions will rotate the selected pattern piece in the direction and by the amount indicated. Use **45°** or **–45°** to rotate pieces for bias grain. Use **90°** or **–90°** for cutting stripes and checks on the opposite grain. Figure 6.22 shows the various positions of the dark blue piece after rotation.

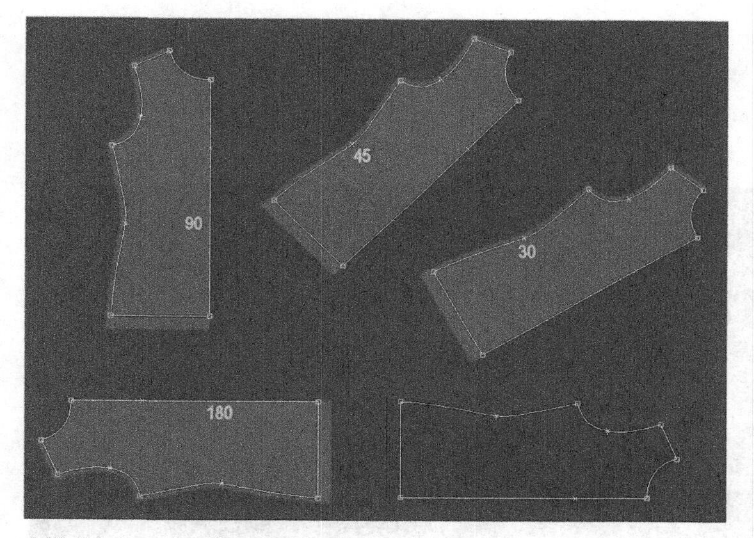

6.22 Rotate a piece by degrees

F2 – Rotate 2 Pts rotates the pattern piece according to any two chosen points. Click on the first point, then the second point and the system will rotate the piece so that the points are on the *x*-axis without distorting the piece. Figure 6.23 shows a half front shaped waistband before and after rotating on the CF line.

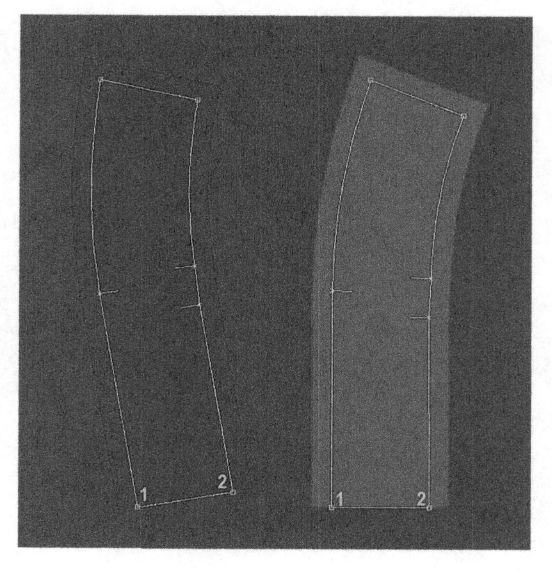

6.23 Rotate 2 points. A shaped waistband realigned on the CF line

Width 50.00 mm Lock
Height 50.00 mm Lock
Rotation Decimal degree Lock
1 2 3

6.24 *Rectangles and squares. Creating a new square pattern piece*

6.3.3 *Geometric shapes*

F2 – Rectangle creates a rectangle or square. There are two choices. Click and drag the cursor until the rectangle is the required size, then click again to complete the action. Or click and drag and enter the values required into the dialogue box. Press the **Enter** key and then left click to complete the action. Good for straight waistbands, pockets, tabs etc. (Fig. 6.24).

Once the rectangle is formed (1) use **F4 – Seam** and click drag in the centre of the rectangle (2) until it turns a luminous blue. Right click to complete the function and a new sheet (3) will be created with the rectangle as a dark blue usable sheet.

F2 – Circle creates a circle. On a new sheet a circle can be created in exactly the same way as the rectangle described above. On an existing sheet it needs an existing point at the position of the centre of the circle. Click on the point then drag until the size required. Or use the **dialogue box** to enter the diameter value.

This is a useful tool for checking measurements, for instance with darts (see Fig. 6.25). Create two circles of the same diameter using the dart opening notches as the centre point. The apex should be positioned where they cross. In Fig. 6.25 the dart apex has been repositioned leaving the purple print line to show the previous place. Used at a dart apex, it allows you to check that the notches at the dart ends are exactly the same distance from the point. Delete the circles after use.

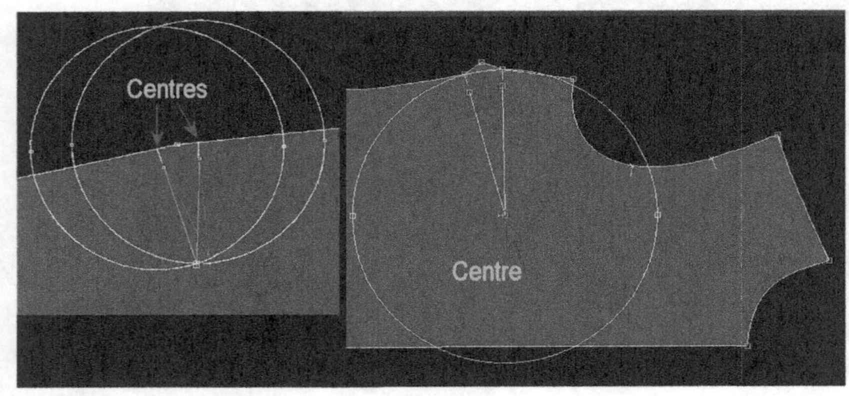

6.25 *Using circles to check dart ends*

6.4 F3: modifying lines and points

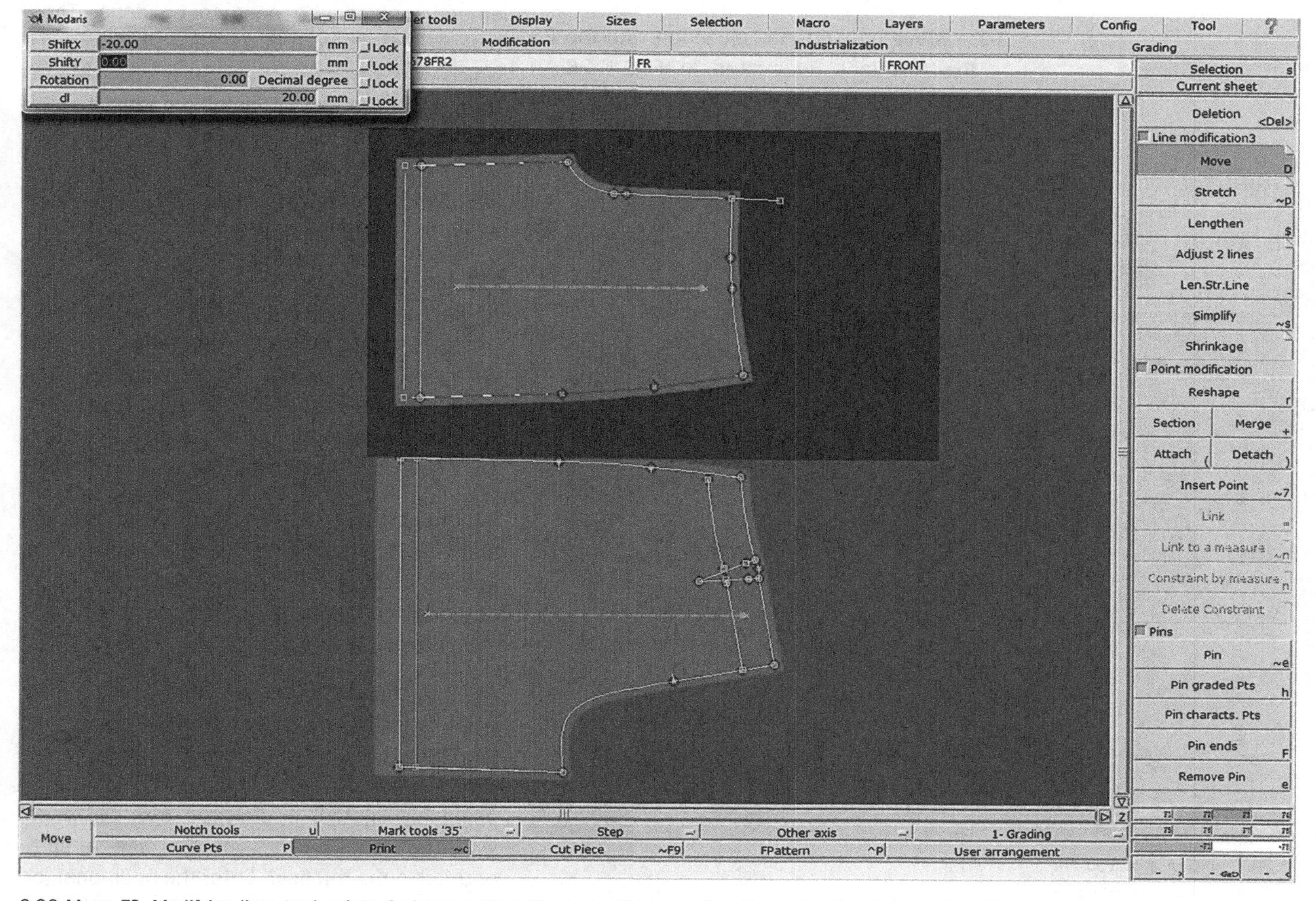

6.26 Menu F3. Modifying lines and points. A shorts pattern illustrates the move function using the dialogue box. Notice the points have red halos indicating that the pin function is selected

6.4.1 Delete

F3 – Deletion is used to delete a point or a line. Simply click on any **characteristic** or **curved** point to be deleted.

To delete a **corner** point, first use **F3 – Merge** then **F3 – Deletion**.

To delete internal lines and their associated points hold **Shift** at the same time. Figure 6.27 illustrates a line which is then deleted without the use of **Shift**, leaving **characteristic points** which will need to be deleted in a second action. The final illustration shows that, with the addition of the **Shift** key, the whole of the line and its points are deleted in one action.

If you try to delete a perimeter line a **prompt box** will appear, warning that it will delete the whole piece. Select **Cancel/Abort**, unless you want to delete the whole thing.

To delete a whole sheet, use **Sheet – Delete** from the upper tool bar. Hold **Shift** while using **Sheet – Delete** to delete the flat pattern (grey sheet) as well (Fig. 6.28).

If you make a mistake use **Ctrl** + **Z** and the piece will be retrieved. However, it is advisable to make a copy immediately and then delete the retrieved piece which has become unstable.

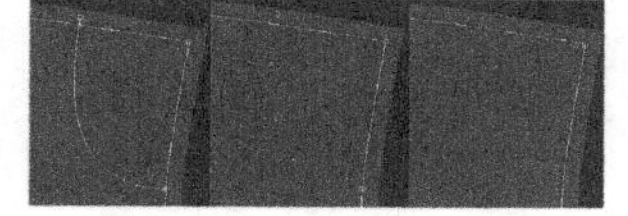

6.27 Effect of deleting a line with and without holding Shift

6.28 Working sheet and grey construction layer

6.4.2 Line modification

F3 – Move is used in conjunction with the **F3 – Pin** functions, illustrated in Fig. 6.26 to lengthen a pair of shorts. (See below for full explanation of **Pin** functions.)

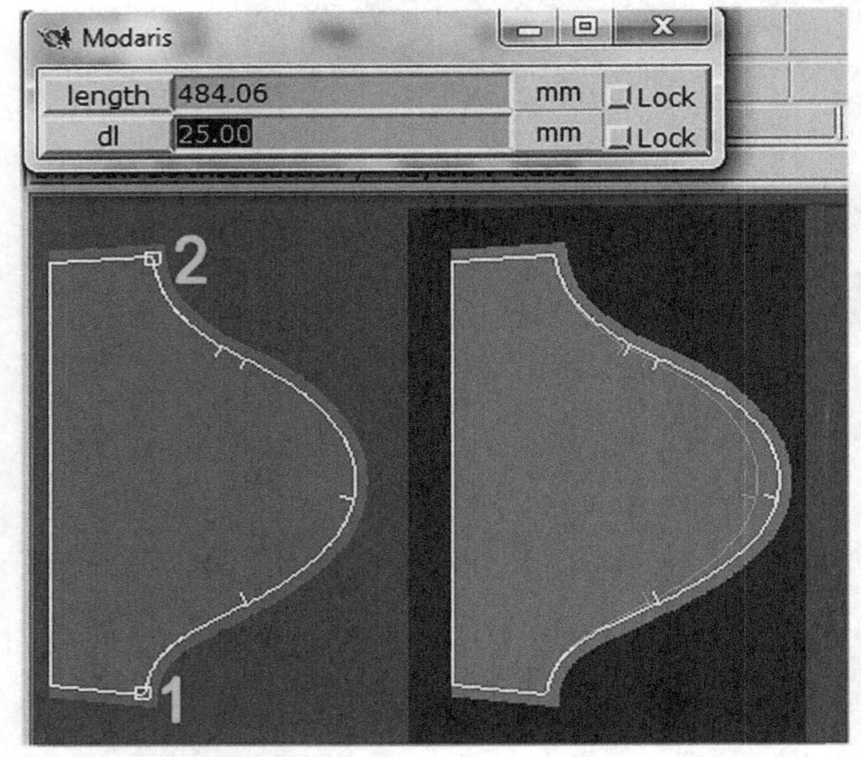

6.29 Lengthen a curved line

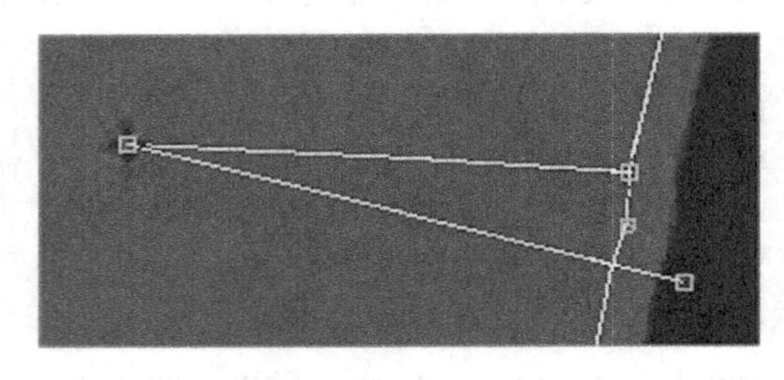

6.30 Adjust 2 lines

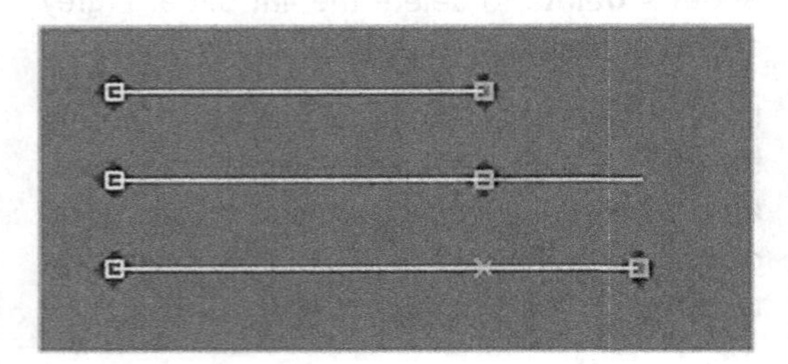

6.31 Lengthen a straight line

First activate the **Print** button so that the amendment can be seen to be done. Then activate one or more of the **F3 – Pin** functions which work simultaneously. Select **F3 – Move**, then select the line or lines (hold **Shift** if more than one) to be moved using the right mouse button. Having made the selection, click on the line and **move** in the desired direction. Click again to complete.

The function can be done freehand, using the on-screen measurement guide, or specific values entered into the **dialogue box**. This is the function to use when lengthening or shortening a series of pattern pieces and note that the **dialogue box** retains the last value entered so it need only be entered once.

F3 – Lengthen is used to lengthen or shorten a curved line between two end points by a specific amount while maintaining the position of those end points. A useful application is for sleeveheads (Fig. 6.29).

Activate the **print** function. Click on the first point of the line, and then the end point. Use the **Space bar** if there is more than one choice. Hold the last click in order to see which line is being highlighted. A **dialogue box** will open with the measurement of the line already entered in the top field. Enter the additional (or minus) amount required in the lower field and click to finish. The line will be amended to achieve the length.

F3 – Adjust 2 lines is used to clip two lines together, and remove any surplus overhang (Fig. 6.30). Click on the line to be kept, and then on the line to be intersected. The first line will be shortened or lengthened accordingly. Repeat in the opposite direction to clip two lines to a neat corner. Do this before extracting a shape as a new pattern piece using **F4 – Seam**.

F3 – Len.Str.Lin (lengthen straight line) simply lengthens an existing straight line. Click and drag on the line until it is extended the required amount (Fig. 6.31).

6.4.3 Point modification

F3 – Reshape is perhaps *the* most commonly used function as this is used to move any point. Click on the point to be moved, and click again to release at new position. Only that specific point will be moved. If the **print** button at the bottom of the screen is activated, a purple shadow line will show the previous position (Fig. 6.32). If you want to return to the previous position, use **Ctrl** + **Z**, to undo the action. Don't forget the **curve pts** button next to **print** on the lower tool bar to see the curved points.

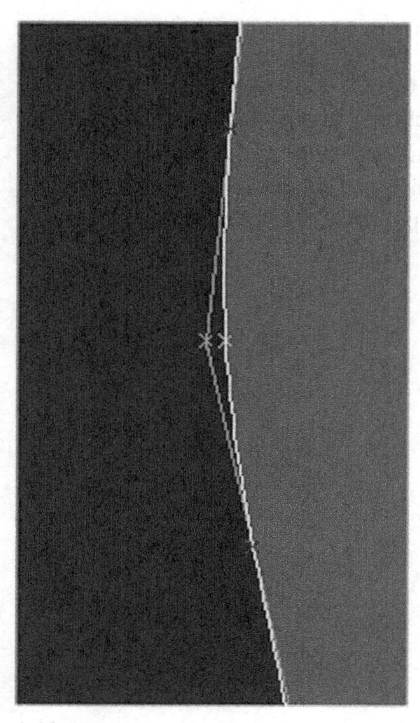

6.32 Moving a point with reshape, with the print function on

Ctrl + **F3 – Reshape** when moving a point will keep it on the true horizontal or vertical. There is the option of blocking the *x*- or *y*-axis by using 0 in the **dialogue box** for better accuracy.

F3 – Section changes a characteristic or curve point into an end point. Click on the point to be modified. When using this function additional points may be automatically added to the line to maintain the shape of the line. This can result in too many points for the line, so activate **print**, and delete as many of the extra points as you can while maintaining the shape of the line.

F3 – Simplify is intended to remove excess curve points (See **F3 – Section** above) but I find it better to delete individual points as required.

F3 – Merge changes a corner point into a characteristic point. Click on the point to be modified. This may affect the seam line values, but is easily remedied, see **F4 – Line seam**.

F3 – Attach joins two points together into a single point. When using the **F4 – Seam** function it is possible that some of the points in the new sheet are not attached to each other fully. This will be noticeable by the red halo around the points. Zoom in close and you will see two points close together but not joined properly (See Fig. 6.33). Click on the first point, then the second point and they will become a single point and lose the red halo.

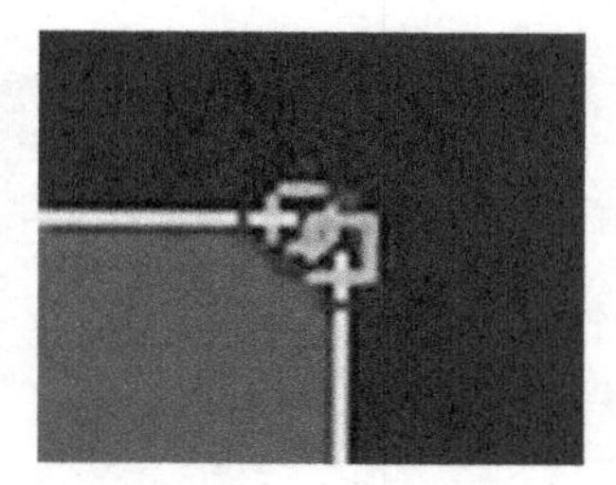

6.33 Close up of non-attached points

F3 – Detach is used to separate one point from another, the reverse of **F3 – Attach** above. Use this function to separate an internal line that has attached itself to the perimeter line. Select **F3 – Detach** and click on the points to be disconnected. This action detaches the point but does not move it. Now use **F3 – Reshape** to actually move the point, using the **Space bar** to scroll through the options and select the required point.

F3 – Insert point is used to attach a **slider point** or **slider notch** to a line, turning it into a characteristic point. Click on the point.

6.4.4 Pin functions

The **Pin** functions are used to hold the pattern piece in place while a moving action is used on a selected section. These functions can all be used simultaneously.

A **pinned point** has a red halo. In Fig. 6.26, the points of the shorts pattern have been **pinned** to allow the **move** function to lengthen at the hem edge. This function can be activated by the keyboard **p** so if you unexpectedly find points have turned red you may have accidentally hit **p**. Activate **F3 – Remove pin**.

F3 – Pin is used to hold individual points in position while others are being moved using the **F3 – Move** function. Note that once multiple points have been pinned using these commands, any single point can be unpinned using the **F3 – Pin** function, which you will need to do if there is a pinned point along the section that you wish to move.

F3 – Pin graded Pts pins all the graded points.

F3 – Pin charact. Pts pins all the characteristic points.

F3 – Pin ends pins all the end points.

F3 – Remove pin removes all the pins. Because the pin functions have keyboard shortcuts it is possible accidentally to activate the pin options, which will be indicated by red circles around the points. Use this function to remove the action.

Chapter 7

A guide to the toolbox in Lectra Modaris pattern cutting software: F4–F8

Abstract: The purpose of this chapter is to introduce the Modaris user to the specific pattern cutting toolbox displayed on the right-hand side of the screen, also known as the trade functions. This chapter focuses on the F4, F5, F6, F7 and F8 menus.

Key words: toolbox, F4, F5, F6, F7, F8, lines and points, notches, Lectra Modaris.

7.1 Introduction

This chapter continues from Chapter 6 by looking at the following functions in the Lectra Modaris toolbox: F4, F5, F6 (incorporating F9 and F10), F7 and F8 (including F12).

7.2 F4: finishing, seam allowances, corners and grain lines

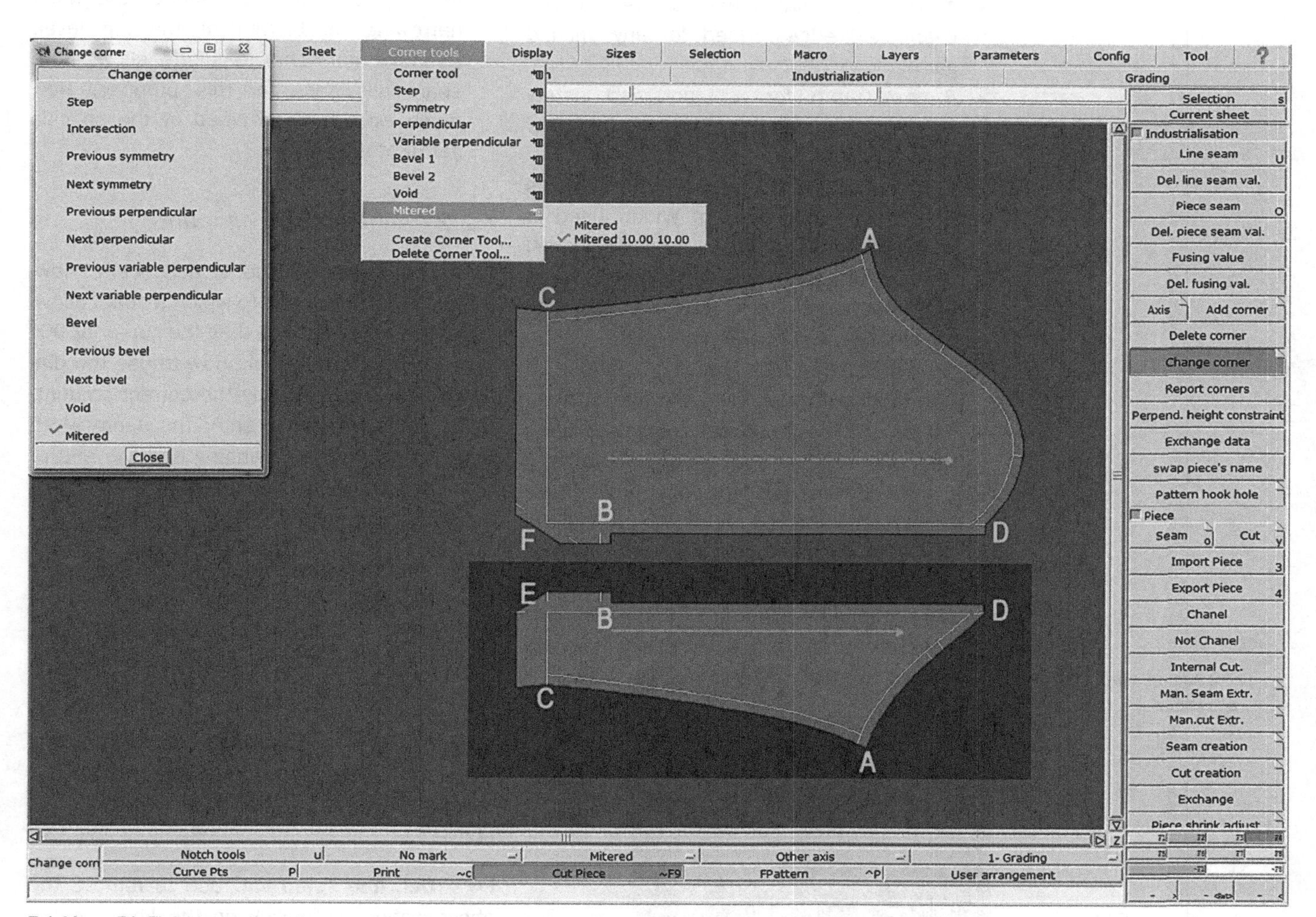

7.1 Menu F4. Finishing, seam allowances, corners and grain lines. A top and under sleeve illustrate most of the corner options, with the change corner tab selected to define the corner shapes

7.2.1 A note on seam allowances

The amount of extra fabric allowed to form the seam allowance varies according to the requirements of the garment and how it is finished. There are some conventions such as using 12 mm (1/2″) for main seams and 6 mm (1/4″) for bagging out seams but they are not always appropriate and depend on three main factors:

- fabric,
- seam finish,
- type of sewing machine.

Guidance for particular situations includes the points below.

- 10 mm for a flat stitched open seam with edges left raw because garment will be lined. Mostly used for tailoring and coats but suitable for any fully lined garment.
- 12 mm for a flat stitch open seam with overlocked edges. Used for any unlined garment.
- 8 mm/10 mm for a four-thread safety stitch machine used for bias-cut and jersey garments.
- 10 mm for a five-thread machine that sews and overlocks in one go. Widely used in many different unlined garment types. Includes trousers and skirts, and often with a small amount of stretch in the cloth. Also for blouses, shirts and night wear.
- 10 mm/12 mm for a French seam used on soft fabrics such as silks and fine cottons.
- 10 mm for a lapped seam typically used on jeans and heavy-weight casual items.
- 5 mm/10 mm for specialist machinery used for cut-and-sew knits, laces and leathers or for adding special trims and need to be assessed individually.
- No allowance – some pile fabrics including sheepskins and faux furs require no extra seam allowance as they are overlocked together in a way which butts the edges together.
- 6 mm for bagging out seams, i.e. those that form a finished edge such as facings or collar and cuff edges.
- 15 mm/25 mm for turned and stitched hems.
- 30 mm hems for skirt and trouser felled hems.
- 40 mm hems for jackets and coats bagged out with a lining.
- Top-stitched seams will need a seam depth that accommodates the decorative stitch. Add 3 mm to the topstitch width.
- Inlay is a term for allowing extra seam allowance for garment adjustment. The centre back seam of men's trousers is the most visible example where a seam allowance of 30 mm is allowed at the waist narrowing to 10 mm at the hip level. Bespoke garments regularly allow extra seam allowance for this purpose. How much extra is determined by the specific garment maker.

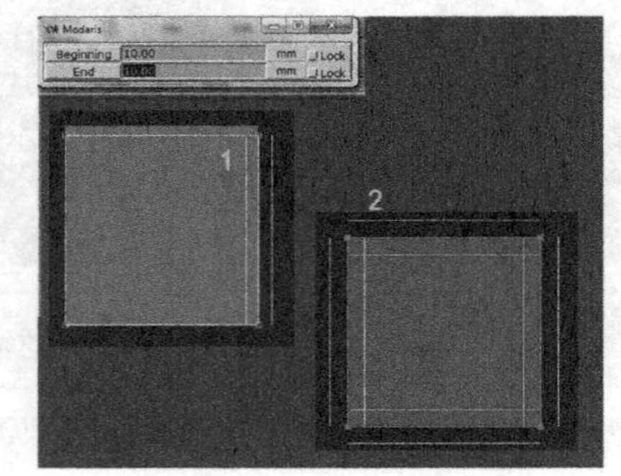

7.2 Adding seam allowance

7.2.2 Adding seam allowance

F4 – Line seam is used to add seam allowances. Click on the line to which a seam allowance is to be added and drag the cursor at 90° to the line. It is advisable now to use the **dialogue box** to add exactly the correct amount. Using the **arrow keys**, enter the seam width value into both the beginning and the end of the line fields. Press the **Enter** key to complete the action (see Fig. 7.2).

More than one seam can be selected at a time. Right click on the first line, then hold **Shift** and select further lines, release **Shift** and enter the values into the **dialogue box**. Press **Enter** to complete the action.

A quick way to add the same value to all the lines on a sheet is to **click and drag** a box around the whole piece. Then continue as above.

F4 – Del. line seam val. Use to remove the above action. Left click on the line and the seam values will disappear.

7.2.3 *A note on grain lines*

The straight of grain is the direction of the warp thread on the fabric and allows the layplanner, be it a person or computer program, to place the pattern pieces correctly onto the cloth. Points to note are outlined below.

- Every pattern piece requires a grain line.
- Be disciplined and keep all of your grain lines orientated the same way.
- Grain lines are generated from the centre back and centre front block positions.
- Grain lines can be repositioned provided it is done with awareness.
- Placing a piece slightly 'off grain' to other pieces will result in the twisting of the garment.
- Try to keep grain lines at a right angle to the waist, hip or hem line.
- The bias grain is at a 45° angle to the grain line.
- Stripes and checks may need to break the rules.
- How the **grain line** is plotted out is determined in JustPrint.

7.2.4 *Axis*

F4 – Axis There are a number of different axes accessed via the corner tab sub-menu. The most frequently used are the **grain line** and **special axis** (see Fig. 7.3).

F4 – Axis – grain line is displayed as a green line. Click at the beginning of the grain line position, then hold the **Ctrl** key, which will keep the line on the true *x*-axis, while dragging the cursor to the end point. Note that the system allows only one **grain line** to a pattern piece.

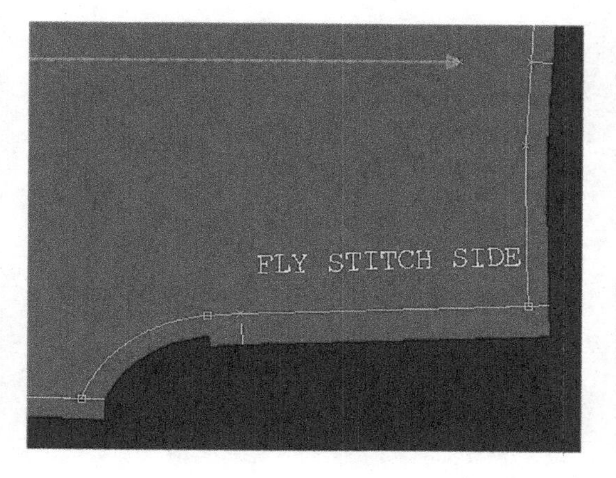

7.3 Axis showing the grain line and special axis text

F4 – Axis – special axis is displayed as a purple line and allows text to be written onto the piece. Create the **special axis** first by clicking at the beginning of the line and then at the end of the line. Then select **Edit – Edit** (upper tool bar) click on the line and a cursor will appear. Make sure it is purple, the same colour as the line, and type the required text. Press **Enter** to complete. To see the text, press **Cut piece** (lower tool bar) and it will become visible. Turn **Cut piece** off again to continue work on the piece.

Multiple **special-axis** can be used on the same piece.

Use **F3 – Reshape** to reposition if necessary.

Use **F1 – Duplicate** to copy to other pieces.

This is not the only way to add text but it is an efficient way to make comments at a specific place on a piece. Make sure that the relevant option is ticked in **JustPrint** if you decide to add text by that method.

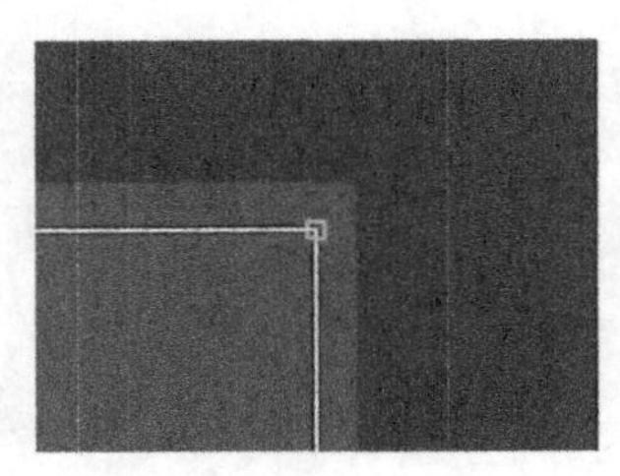

7.4 Intersection corner

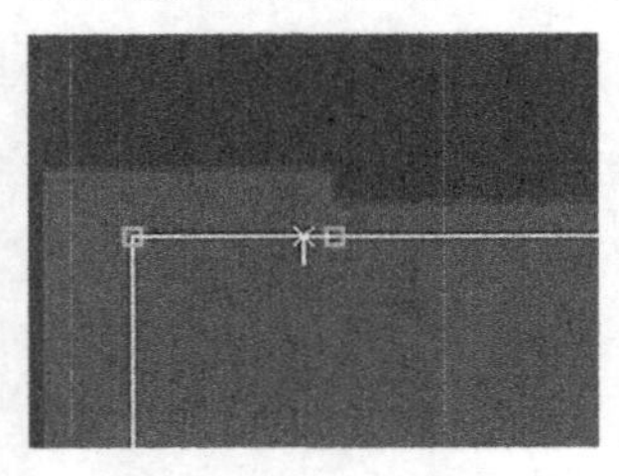

7.5 Stepped seam allowance

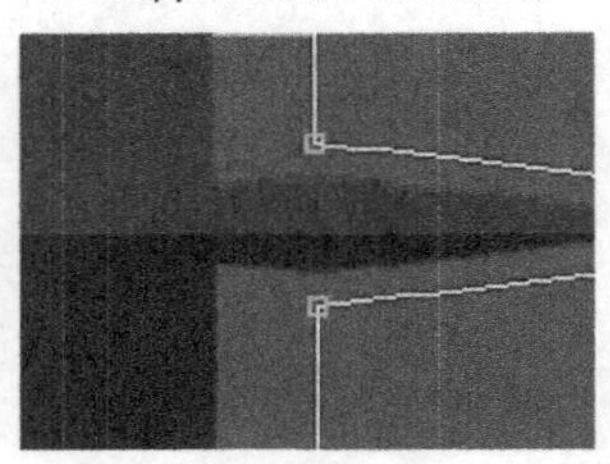

7.6 Symmetrical corner

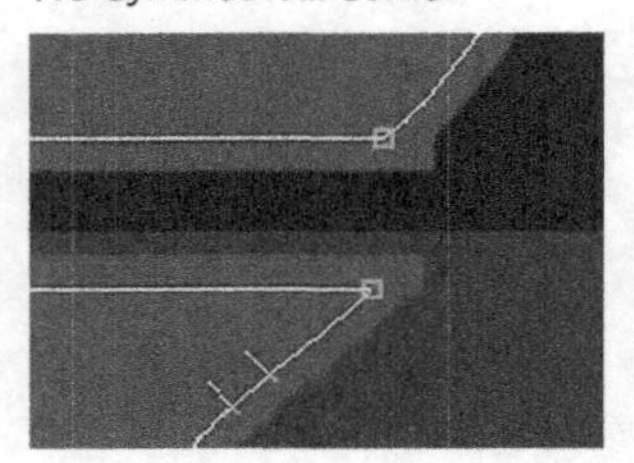

7.7 Perpendicular corner

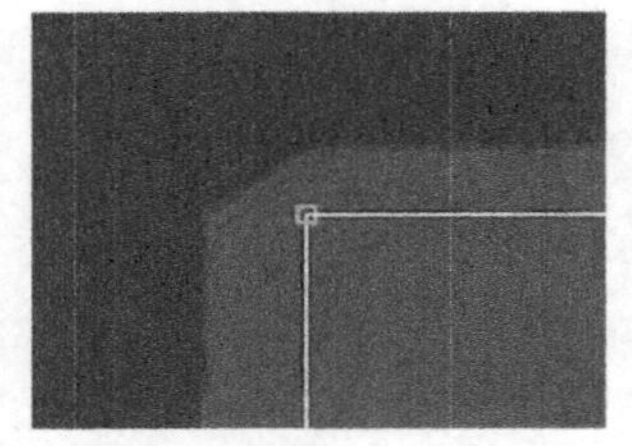

7.8 Bevel corner

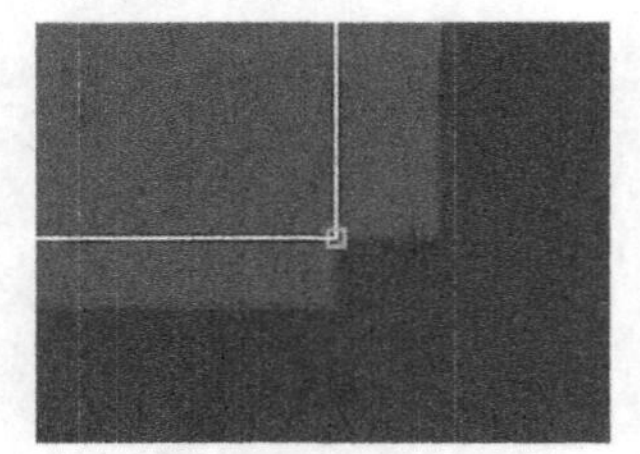

7.9 Void corner

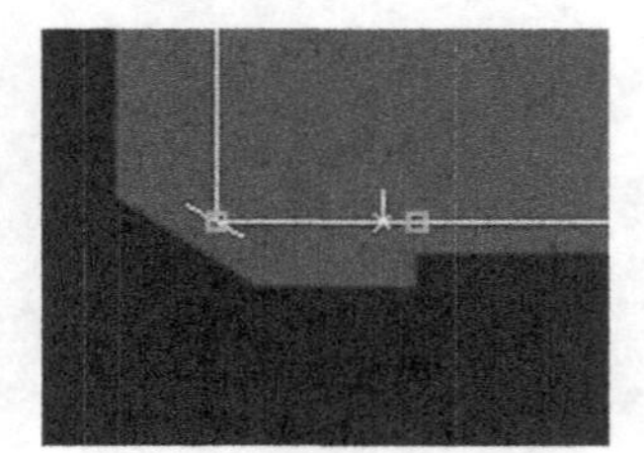

7.10 Mitre corner

7.2.5 A note on corners

Corners, like **notches**, inform the Machinist of the intended making-up order of the garment. Please note the following:

- matching seam ends (corners) ensures seams are correctly aligned,
- corner angles indicate in which direction a seam is intended to be folded,
- correct corner shapes at hem edges ensure a neat turn back.

7.2.6 Corner types

F4 – Change corner is used to vary the shape of the corner intersection according to the requirements of the seam ends when sewing. There is a sub-menu with a selection of options accessed by the corner tab. All the options are useful in different situations. Essentially the function will reshape the ends of the seams according to the type of corner chosen. See Fig. 7.1 for typical uses on a sleeve pattern. Note that **Change corner** should be used in general on end points. The only time **Add corner** is used is to create a 'corner' on a characteristic point. Using **Add corner** several times on an end point might cause overlapped corners, leading to further problems on the piece.

Intersection is the default corner created by the system. Note that not all corners require a special corner, and this is suitable for many positions. Use **intersection** to undo a modified corner (see Fig. 7.4). See also the underarm corners (A) on Fig. 7.1.

Step creates a sharp step where adjoining seam allowances have different values. Here it is not positioned at the corner but along a straight section of the seam at the point of seam value change (see Fig. 7.5). See also the sleeve opening seam section (B) on Fig. 7.1.

Previous symmetry/next symmetry alters the end to be angled to create a symmetrical return (see Fig. 7.6). See the panel seam corners at the hem edge (C) on Fig. 7.1.

Previous perpendicular/next perpendicular alters the pointed seam end to a squared-off end according to the seam direction chosen (see Fig. 7.7). See the panel seam corners where they meet the sleevehead (D) on Fig. 7.1.

Bevel/previous bevel/next bevel cuts a blunt end according to the seam line values, removing excess fabric at the corner (Fig. 7.8). See the underarm panel seam hem (E) on Fig. 7.1.

Void removes the excess from the square end (Fig. 7.9).

Mitered will remove excess from the corner (Fig. 7.10). See the topsleeve, panel seam hem (F) on Fig. 7.1. However, the mitred corner needs some further action, as the default setting will not provide seam allowance. See Chapter 8, Section 8.5.1. 'The standard mitre corner' on how to customise corners.

7.2.7 Piece modification

F4 – Exchange data The place to use this function is after the pieces have been named and the **variant** created. If you then have to amend or change a piece, first carry out the work on a copied or new sheet. Figure 6.23 shows two front waistbands. One is included in the variant and displayed dark blue, the other a copy with amendments (in this case rotated on 2 points) shown light blue.

When the new piece is ready to be put into the **variant** and the previous piece removed, select **F4 – Exchange data.** Click on the original, dark blue piece first and then on the new piece. The information and associated action (such as inclusion in the **variant**) will be changed from the one to the other. Pieces included in the **variant** will be a dark blue provided **orphan pieces** has been ticked in the **Display** drop-down menu (see Fig. 7.11).

F4 – Seam actually has little do with seams as such. This function is used to extract a shape from one sheet and automatically create a new sheet containing that shape, more like tracing off a new pattern piece.

To begin make sure all of the lines intersect each other fully by using **F3 – adjust 2lines.** Select **F4 – Seam**, and click and drag within the new shape and it will turn a luminous blue. Right click to complete the action (see Fig. 7.12).

In Modaris V5, to extract a shape created by the joining or overlapping of two pattern pieces it is necessary to use **F8 – Assemble** (see Fig. 7.13). This function is no longer available in Modaris V6. As an alternative process, once the pieces have been assembled use **Sheet – Copy** to create a flat pattern of the marriage. The extraction with seam can then be performed on the flat pattern.

7.11 Display menu selecting orphan pieces

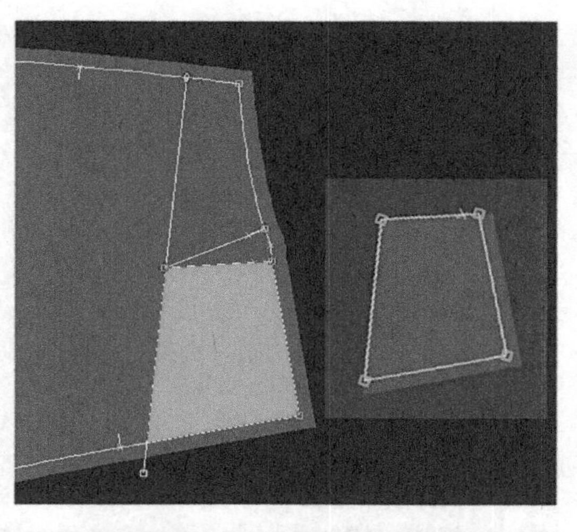

7.12 Seam function extracting a new shape from an existing pattern piece

7.13 Seam function creating a new pattern piece from two separate pieces

7.2.8 Exchange seam position

F4 – Exchange is not to be confused with the **F4 – Exchange data** above. This function controls the placing of the seam allowance in relation to the solid white line on screen. By default, when the piece is first **digitised** or created on screen, the perimeter line will be the **cut line** of the piece, as shown by pattern piece-A in Fig. 7.14. This means that when a seam allowance is later defined by the **F4 – Line seam** function it will be towards the interior of the piece. If the digitised pattern has had the seam allowance included, as pattern piece-B in Fig. 7.14, then this is fine. Note, though, that if seam allowances are added to the pattern before digitisation, as many companies do, seam allowance will not be defined on screen at all. See Chapter 4 for more discussion on this point.

However, if the seam allowances are to be added as additional to the digitised line, that is added to the outside of the white line, pattern piece-C in Fig. 7.14, then the line must be **exchanged**. To change the placing of the allowance to the other side of the cut line, use **F4 – Exchange**, and click in the centre of the piece. Individual lines can be **exchanged** but it is advisable to select all lines. Use only one method throughout a specific pattern (see Fig. 7.14).

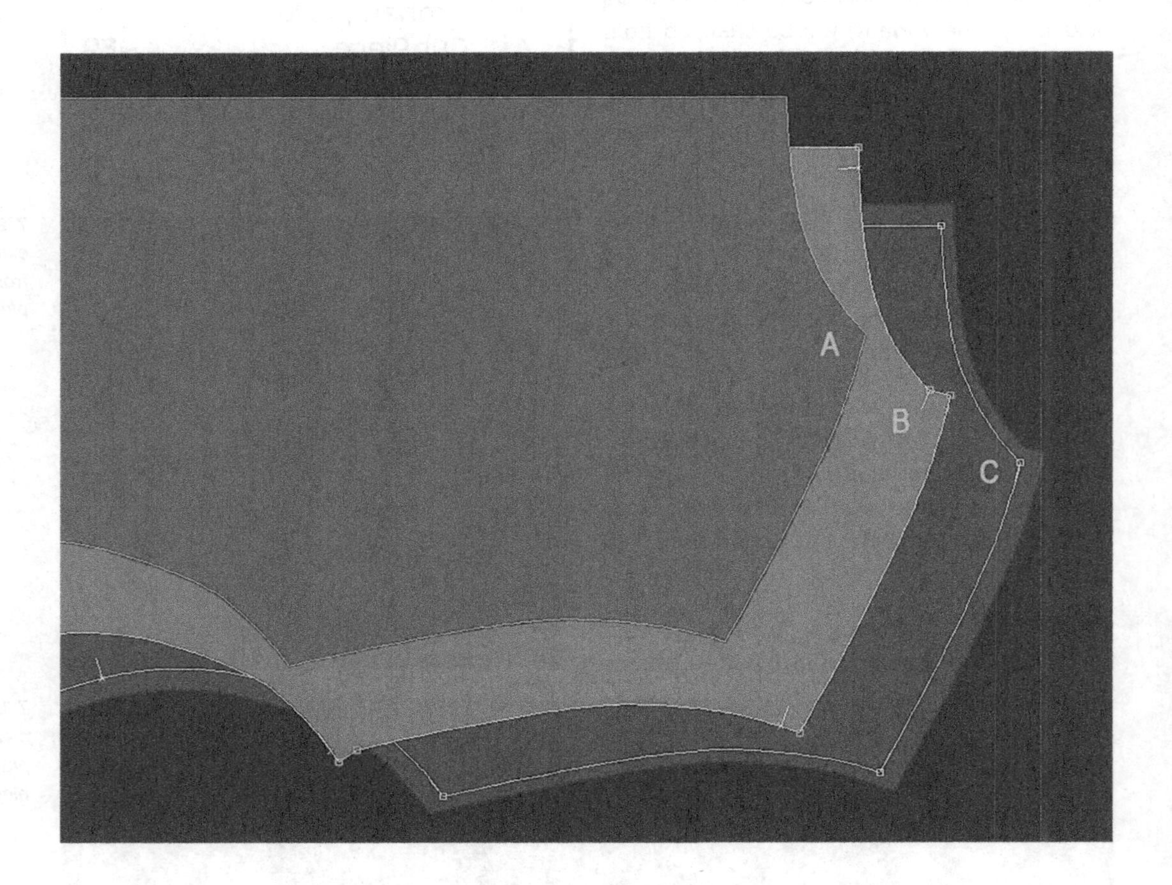

7.14 Exchange seam. Three variations of the same pattern piece

7.3 F5: derived pieces

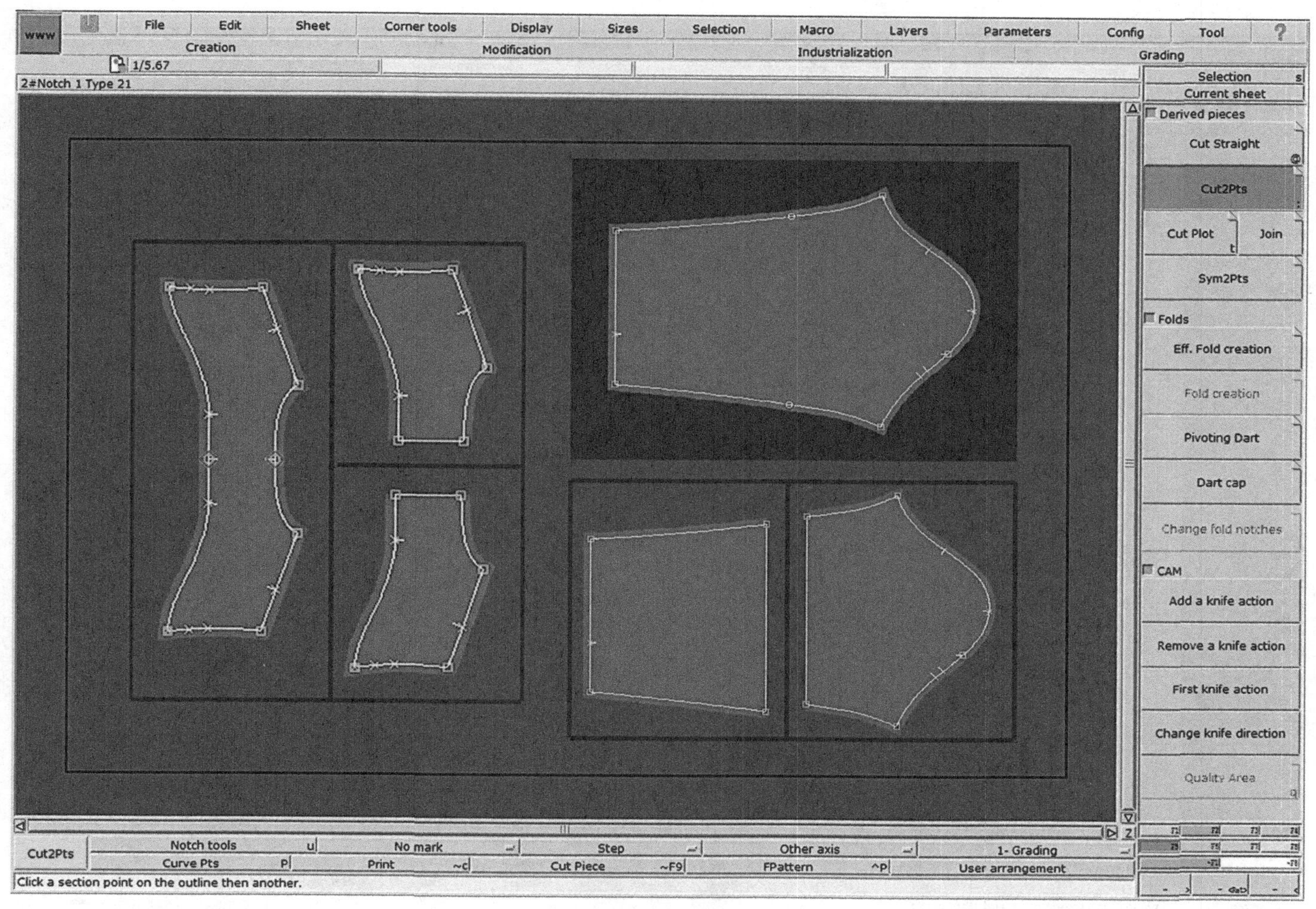

7.15 Menu F5. Derived pieces. A yoke and sleeve show two variations of Cut2Pts

F5 – Cut2Pts will split a piece into two separate pieces on new sheets along a chosen straight line. Click on the first point, then the second point. Two new sheets will be created leaving the original intact. This doesn't have to be a CF or CB, although is very useful for splitting a mirrored piece, but can be any piece. The function can only be actioned between two perimeter points and will give a straight edge to the resulting pieces.

F5 – Sym2Pts does the reverse. Click on the first point (a) of the line of symmetry, then (b) the other end. A new sheet will be created with the open or mirrored piece (see Fig. 7.16).

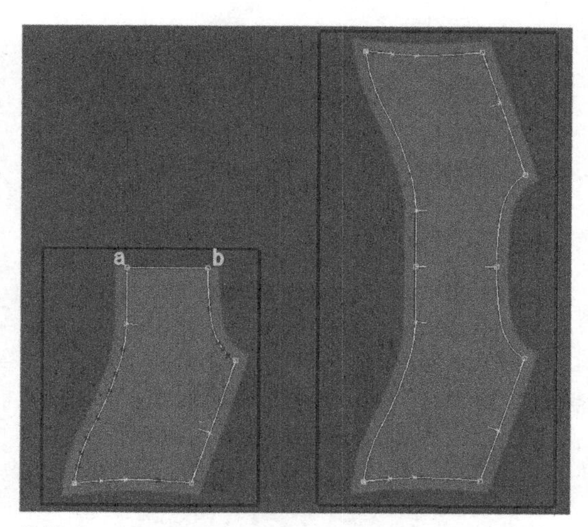

7.16 Sym2Pts, showing a yoke mirrored along the CB line

7.4 F6: grading

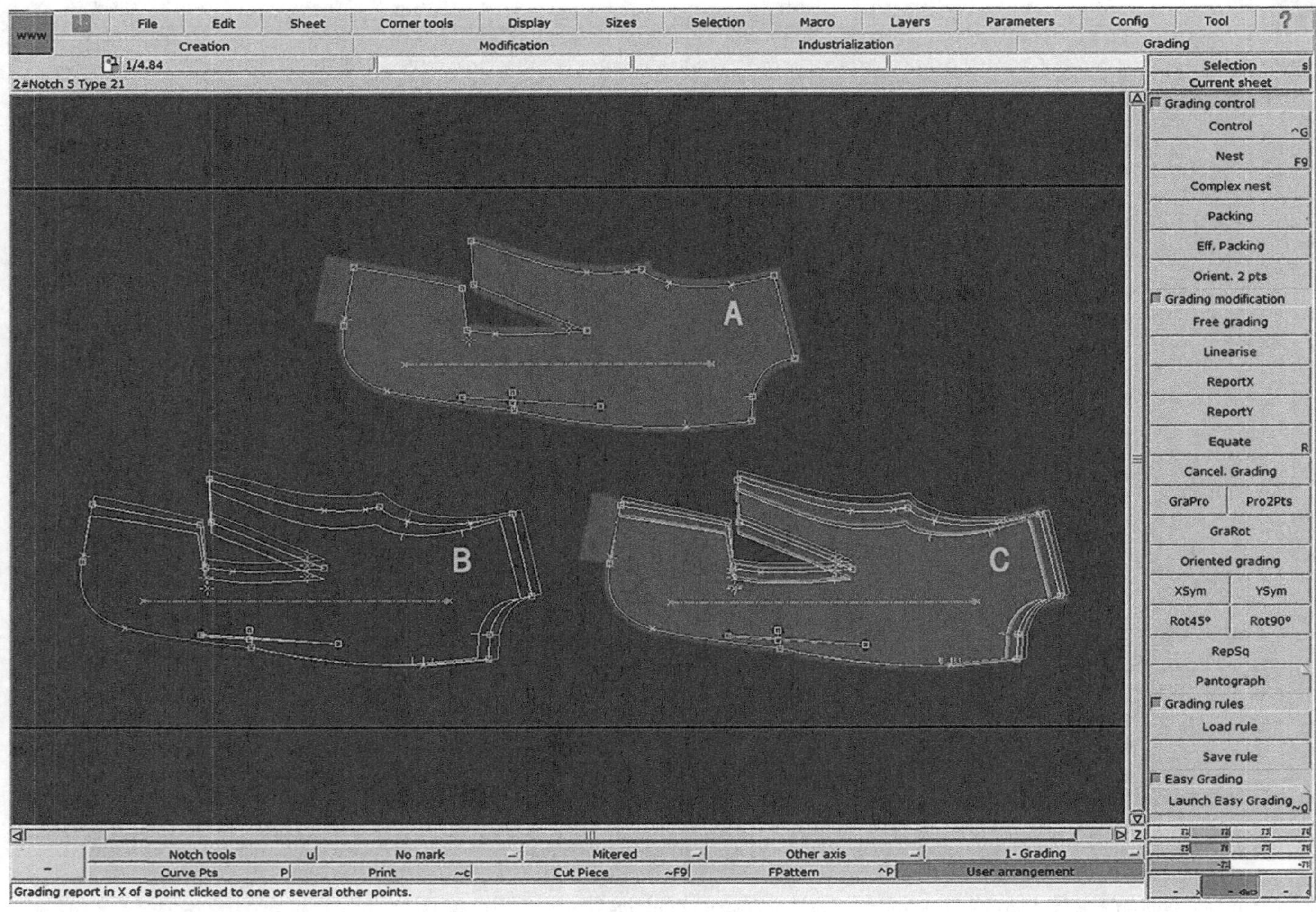

7.17 Menu F6. Grading. A jacket front illustrates the nest display options

The grading menu is, obviously, totally concerned with grading. Here it is sufficient to show where those functions are located. To see if a pattern is already graded, press **F9**. Please note:

- If the contour lines turn only orange, (A) with no nest of sizes visible, the pattern is ungraded. Press **F10** to return to the normal setting.
- A graded pattern will display a nest of sizes, (B) the base size as a white line, the largest as an orange line and the smallest as a yellow line. Pattern modification can result in grading distortion so never assume a pattern is graded correctly without a Pattern Grader's approval.
- To see a complete size nest (C) press **F12** immediately prior to selecting **F9**. When **F12** is selected the whole size range displayed in the text box on the left will become highlighted. Each size in the nest will be displayed in a different colour.

7.5 F7: evolution system

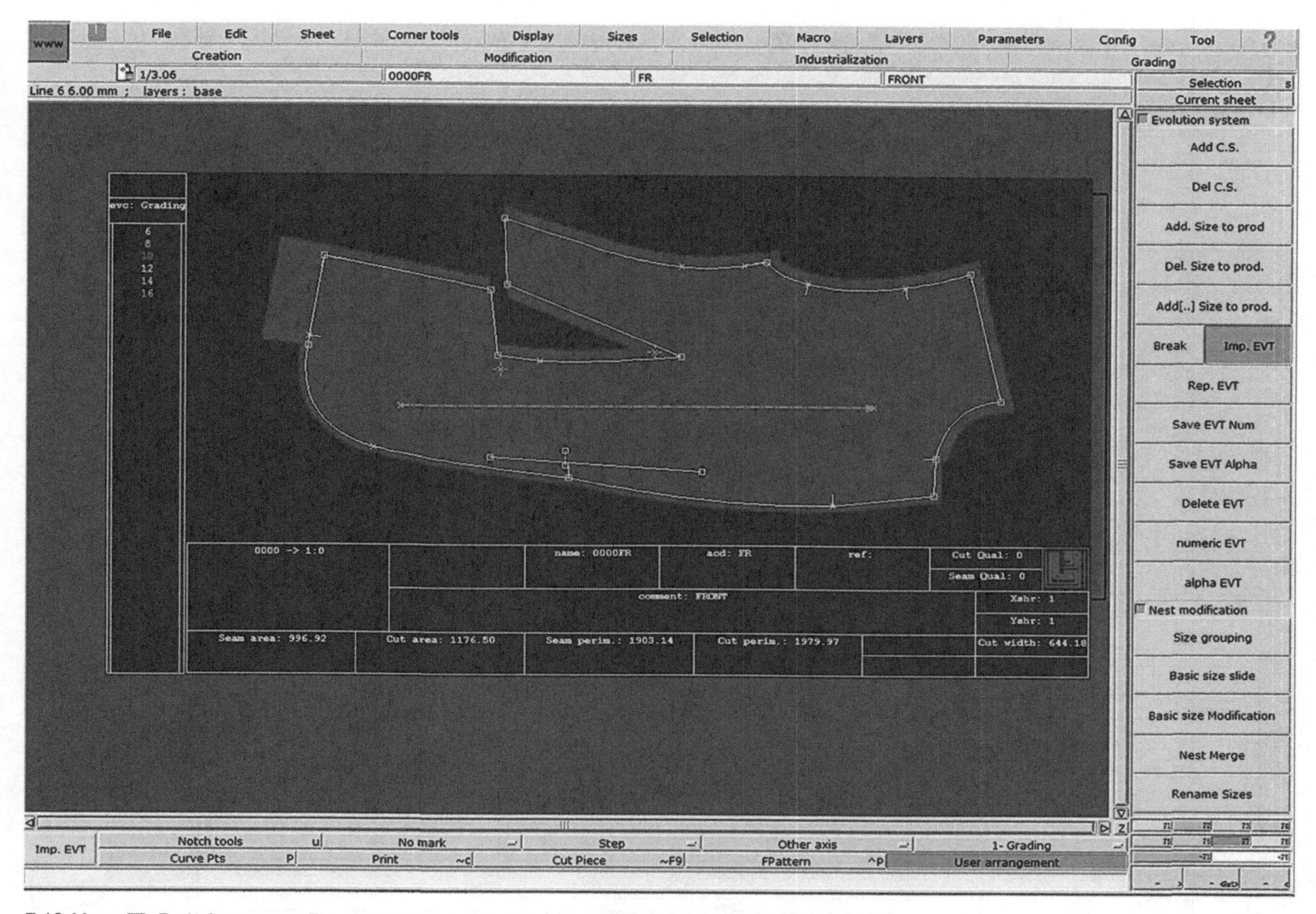

7.18 Menu F7. Evolution system. The title blocks on the left are inserted by the use of ImpEVT

This menu is all about sizes, their designation and modification. Most of the functions are for advanced use and in conjunction with the grading functions. One, however, is essential.

F7 – ImpEVT stands for import evolution. The first action when creating a pattern, is to associate a size range. Select **ImpEVT** (1) and double click in the centre of the title page. A new window will open for step 2 where you select the relevant sizing code: 6–16, SML, 36–44 etc. The size range (3) will now be inserted into the left-hand title block (see Fig. 7.19). See Chapter 11 for how to create size files.

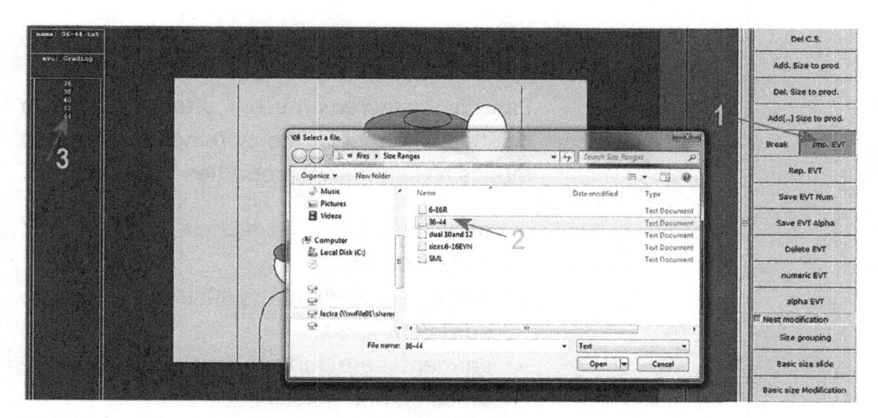

7.19 Various size range files accessed by ImpEVT

7.6 F8: measuring and assembly

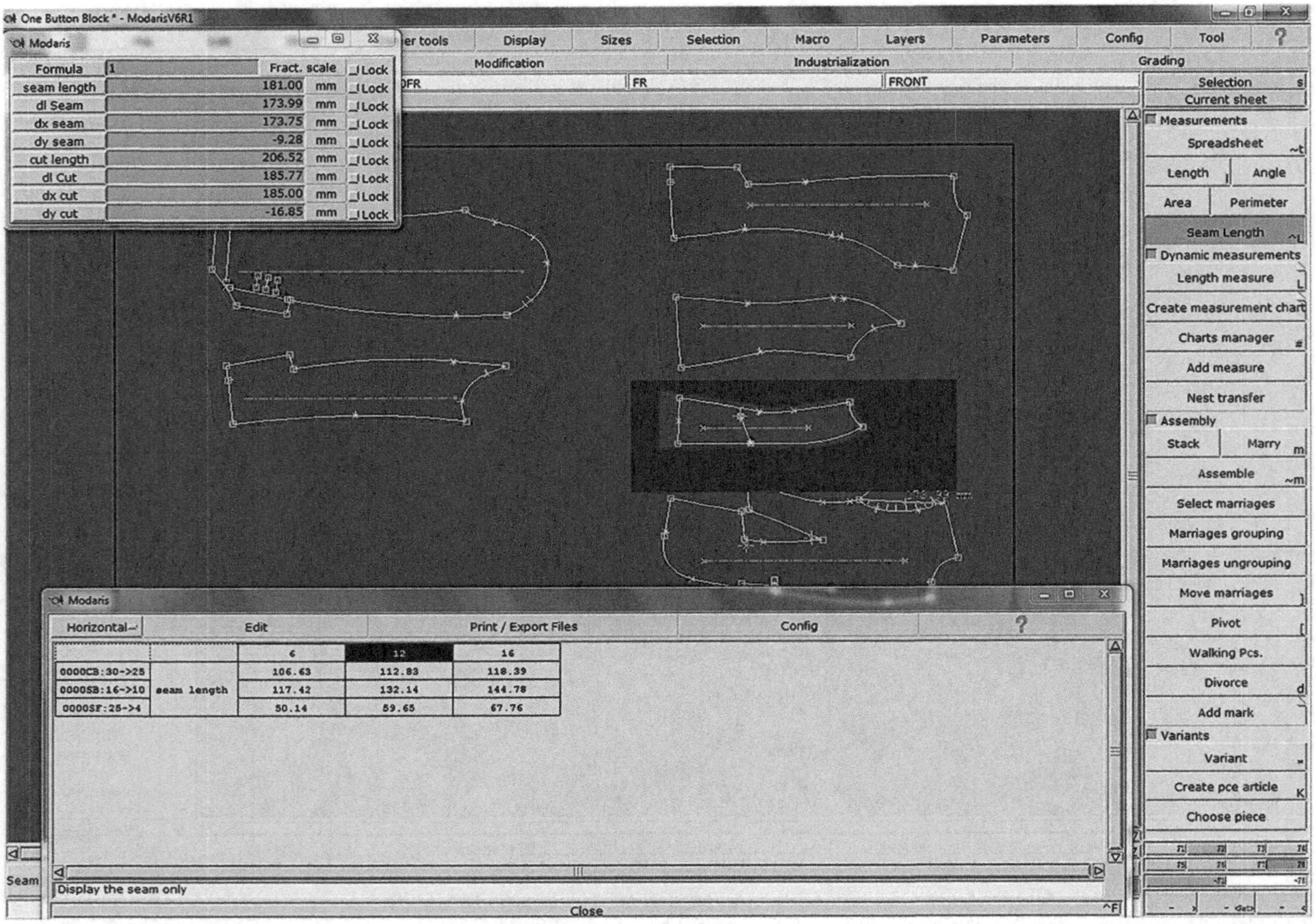

7.20 Menu F8. Measuring and assembly. Jacket pattern with spreadsheet and dialogue box

7.6.1 A note on measuring

Pattern measurements are often different to finished garment measurements. Various fabrics have different properties but in general:

- once cut fabric stretches, shrinks and moves off grain,
- fabric is eased and gathered during the sewing process,
- garments are further manipulated during pressing and finishing.

Measuring needs to take account of these factors and allow extra fabric as appropriate.

The **spreadsheet** in this menu allows a series of measurements to be taken, added up and compared and is a very valuable on-screen function. Figure 7.20 shows a jacket pattern placed to measure the armhole and compare it to the sleevehead. The separate sections of armhole will be measured first and added up, and then the sleevehead sections in the same way. The amount of ease allowed can then be reviewed and adjusted as necessary.

A **dialogue box** will open for each individual measurement taken, and show the various measurements of the line. Hold the left mouse key down and this can just be read on screen as needed without opening the spreadsheet. Release to complete and the **dialogue box** will close automatically.

The **spreadsheet** will store any measurements taken regardless of being opened or not.

The **spreadsheet** display will depend on the selection made by the user under the **Config** menu within the spreadsheet box.

7.6.2 *Measuring*

F8 – Spreadsheet opens the spreadsheet displaying the measurements acquired on screen with either **F8 – Length** or **F8 – Seam length**.

F8 – Length measures the distance between any two points, and not just on a seam line. Click on the first point, then on the second point. The measurement can be seen on screen but will also be entered into the spreadsheet (see Fig. 7.21).

F8 – Seam length is used to measure seam lengths. On screen the direct measurement between the two points will be displayed, shown by the white line and arrow. All the related seam measurement will be entered into the spreadsheet. Figure 7.22 shows all of the options available.

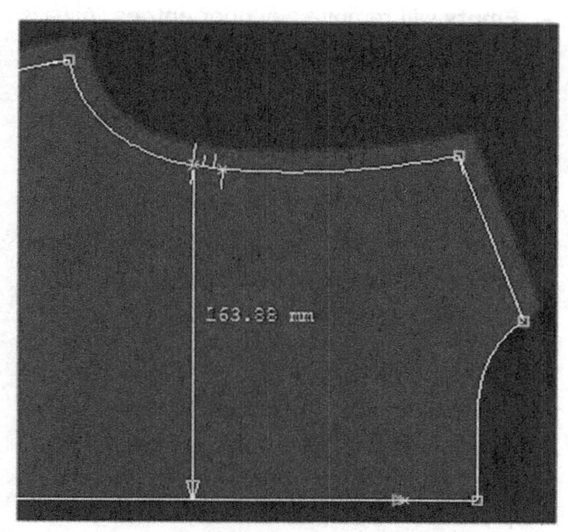

7.21 Length measurement showing the direct measurement

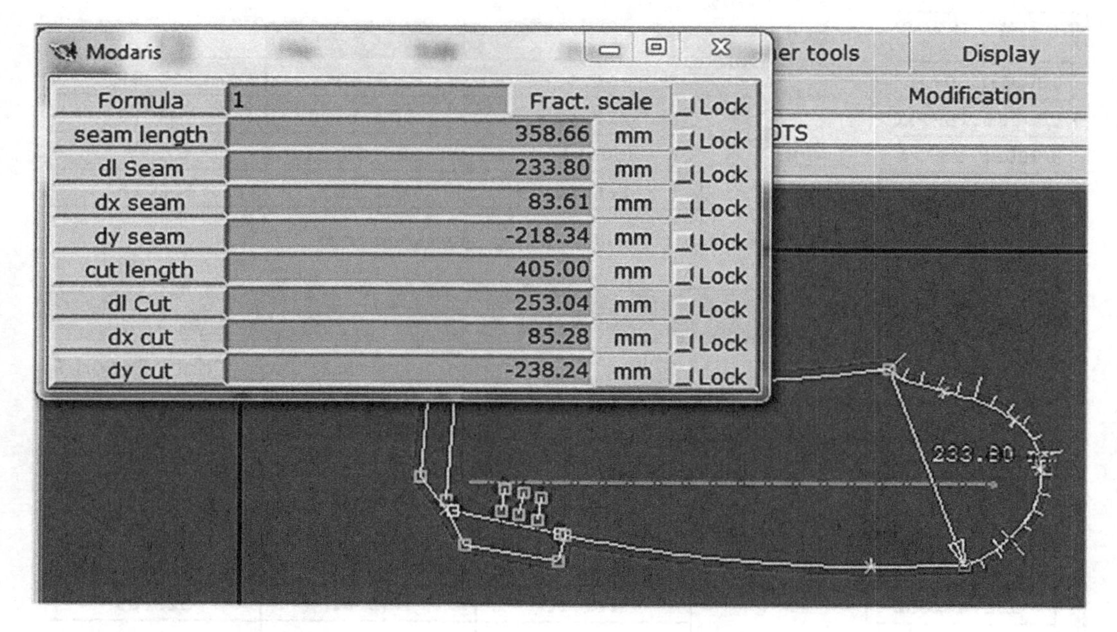

7.22 Seam length measurement dialogue box displaying all the related measurements of the sleeve head

Figure 7.22 shows the seam length measurement dialogue box displaying all the related measurements of the sleeve head. Figure 7.23 displays the seam length only of the base size 12, along with the smallest, size 6 and largest size 16.

Under the **Edit** menu within the spreadsheet there are a number of further actions.

- **Empty** will remove any prior entries. Always do this before beginning a measurement comparison exercise.
- **Cumul** (cumulative) will add up the previously taken measurements.

The **spreadsheet** can then add up a series of measurements, or the information cleared. This is useful for adding up armhole measurements and comparing with the sleeve head.

The left-hand column of Fig. 7.24 identifies the pattern piece and the points between which the measurement was taken. The numbers after the colon refer to the specific points on the piece, so that if you amend the pattern and retake the measurement using the same points the spreadsheet will update.

You can see here that the pattern number is 0000 followed by the piece identifier, CB for centre back. SB for side back, SF for side front, and FR for front. A reminder that disciplined labelling in the title blocks ensures ease of identification in other functions.

In the same column the next line shows the cumulation or total for the armhole, which in this case is 485.62 mm. Below again are the TS-topsleeve and US-undersleeve sleeve head measurements totalling 521.64 mm. This gives 36.02 mm of ease in this particular sleevehead, which, by the way is a lot, but not for a wool jacket as in this case.

Use **F3 – Lengthen** to adjust sleevehead ease if required.

Press **F12** immediately prior to taking the first measurement to display all of the sizes in a range. Good for checking grading. See Chapter 8, Sections 8.8.3 and 8.8.4.

Edit

Edit E
Mes. Deletion
Cumul
~V
Empty

		6	12	16
0000CB:30->25		106.63	112.83	118.39
0000SB:16->10		117.42	132.14	144.78
0000SF:25->4		50.14	59.65	67.76
0000FR:66->10		168.12	181.00	192.15

7.23 Spreadsheet detail of edit menu showing 'cumul' for adding up measurements

Modaris

Horizontal | Edit | Print / Export Fil

		6	12	16
0000CB:30->25	seam length	106.63	112.83	118.39
0000SB:16->10		117.42	132.14	144.78
0000SF:25->4		50.14	59.65	67.76
0000FR:66->10		168.12	181.00	192.15
cumulation0		442.30	485.62	523.09
0000TS:11->25		335.87	358.66	376.92
0000US:16->5		145.08	162.98	178.30
cumulation1		480.95	521.64	555.21

7.24 Spreadsheet detail showing cumulations of armhole and sleevehead sections

7.6.3 Assembly

F8 – Marry allows one or more pieces to be added onto a host sheet so that pattern pieces can be checked, by matching seams or **walking pieces** together along seam lines. Click on a point of the guest sheet first, then on the point to be aligned with on the host sheet. The first sheet will then be placed onto the host sheet where it can be further moved and rotated.

Shift + F8 – Marry will flip the piece over during the action. The guest sheets turn a different colour, and each additional added sheet will be a different colour. Multiple sheets can be added to a host sheet. Good for checking hem runs etc. (see Fig. 7.25).

F8 – Assemble is similar to **F8 – Marry**, but requires the selection of two points on the first sheet that will correspond to two points on the second sheet. The first piece will then be rotated and placed accurately onto the second. This is the function to use in Modaris V5 when creating a new piece from the overlapping or joining of two sheets such as creating facings that cross one or more seams, see **F4 – Seam** (see Fig. 7.26). This option is no longer available on Modaris V6.

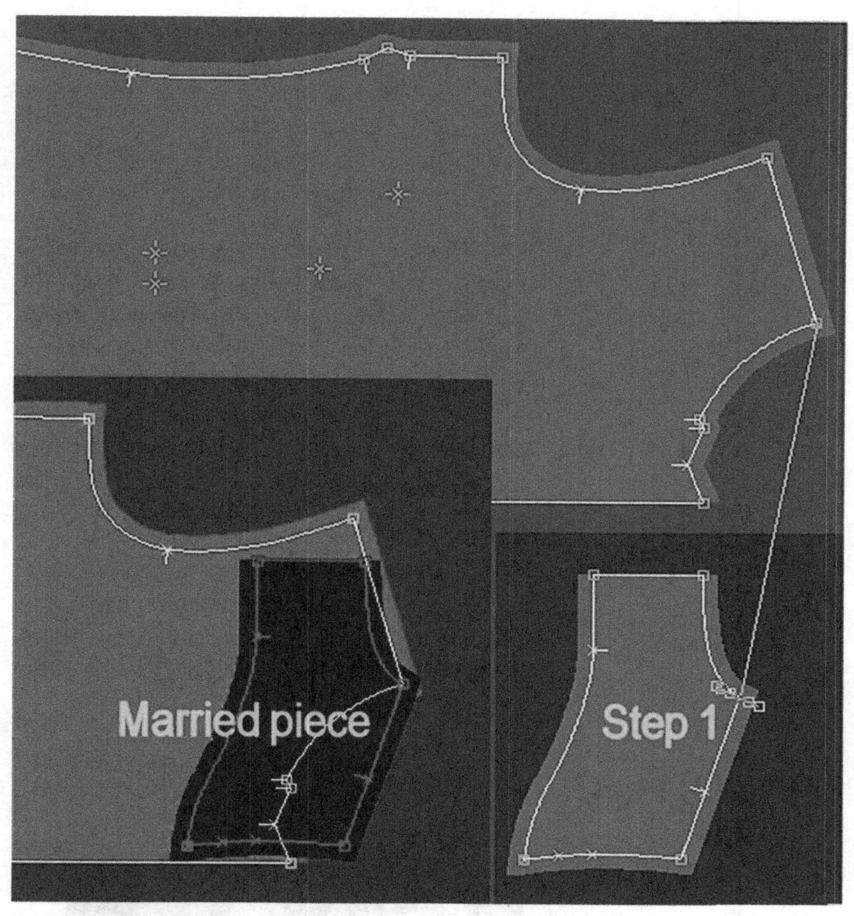

7.25 Marry two pieces

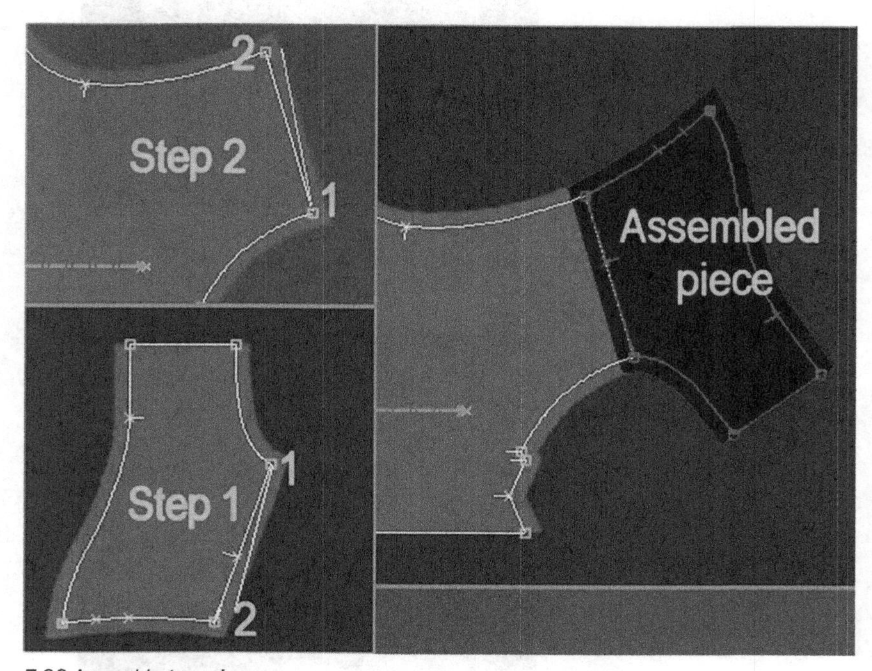

7.26 Assemble two pieces

F8 – Move marriage is used once pieces are married to move the guest piece about on the host sheet. Simply click on a point of the guest piece, coloured red, and move. In Fig. 7.27 the new position is shown as a white outline, but once clicked into place, it will turn red.

F8 – Pivot is used once pieces are married to then rotate or pivot the guest piece about the last selected point. Use **F8 – Move marriage** to reposition the guest piece to the required pivot point. Then click on the piece to be rotated, and pivot as required. Note, if the piece is not pivoting as you require, go back to **F8 – Move marriage** and reselect the pivot point (see Fig. 7.28).

F8 – Walking pieces allows married pieces to be accurately joined on a common seam line, and then rolled, or walked, along the length, stopping and starting as required by use of the cursor (see Fig. 7.29).

Use **F8 – Move marriage** to reposition the guest piece to the required start position. Then click and roll the cursor along the seam line. **Pivot** might also be needed since the lines to be walked need to be as close as possible in order for the tool to work properly. If there is a choice of guest pieces use the **Space bar** to move through the options. The movement of the cursor determines the direction of the walking.

F8 – Divorce is used to separate married pieces. Simply click on the piece to be returned to its own sheet. A useful procedure, when needing to copy a shape or line, is to **marry** a piece as required, activate the **print** function and then **divorce**. A shadow of the first piece will remain on the host sheet (see Fig. 7.30).

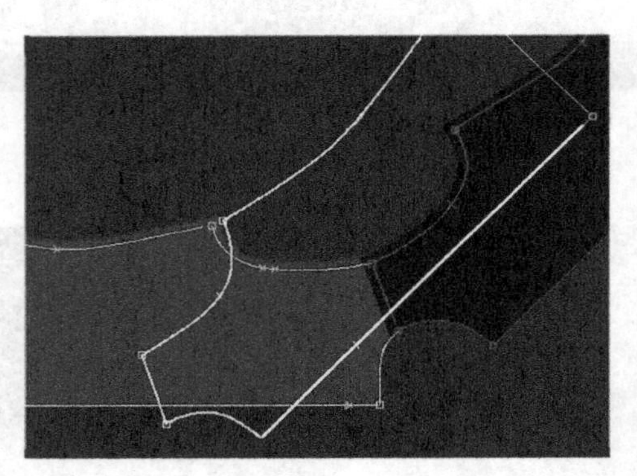

7.27 Move marriage, moving from a shoulder alignment to the underarm point

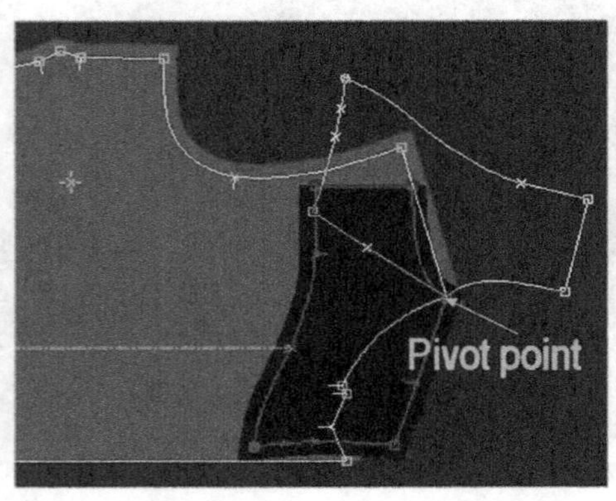

7.28 Pivoting a yoke about a neck point

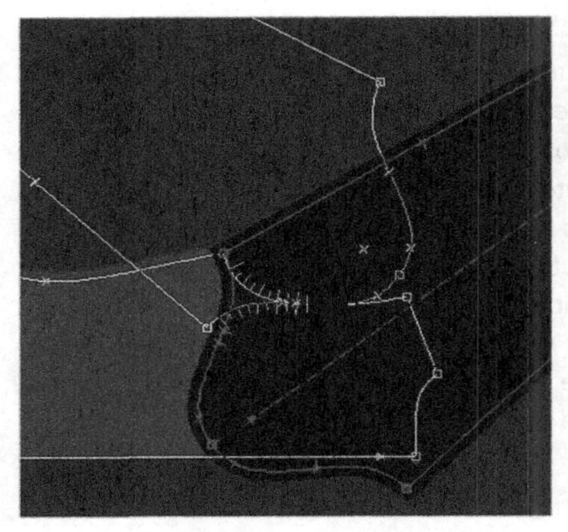

7.29 Walking a sleeve head around an armhole

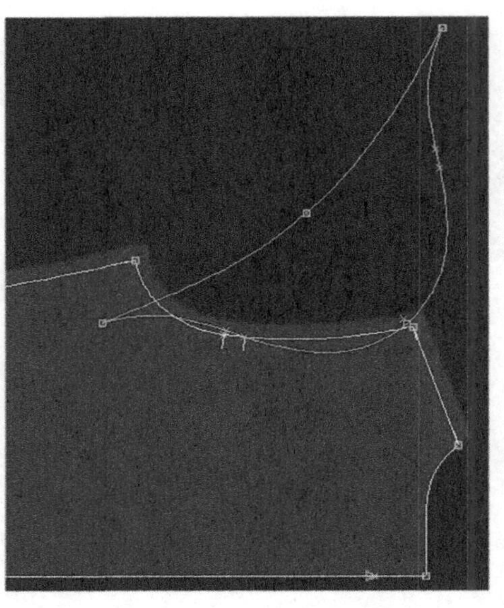

7.30 Print function used with divorce to leave a trace of the pattern piece

7.6.4 Variant

See Chapter 3, Section 3.5 for a detailed look at the **variant**. A **variant** is a spreadsheet in which all of the required pattern pieces are brought together to form a list of the complete pattern. It is here that cutting instructions are made and linked to **Diamino**, the marking making program.

More than one **variant** can be created within a pattern allowing for style variations while utilising many common pieces. For example, various collar or cuff types can be used with the same bodice and sleeve pieces. I would recommend that you use exactly the same numbering as the **model name** for the **variant** as it keeps things simple. But if there are multiple variants they will need identifiers. Suffixes such as LS for long sleeve or SS for short sleeve are useful. It helps to be consistent throughout the system so choose and implement a coding and be consistent in its use. The value of this will become apparent when searching. Chapter 3, Section 3.2 on title blocks explains the links in detail.

F8 – Variant is used to create a new **variant** or open an existing **variant**. Select **F8 – Variant** and a dialogue box will open where you enter the **variant** name. Press **Enter** and a yellow sheet – the **variant** – will appear. This may be off screen so select **J/j** to bring into view. This can be minimised while pieces are selected. Information is automatically saved and needs no further action. To open an existing **variant**, select **F8 – Variant** and, ignoring the dialogue box, click directly on the centre of the existing yellow sheet. The **variant** spreadsheet will open.

F8 – Create piece article is used to select individual sheets for inclusion in the variant. Click once in the centre of the chosen piece and, providing orphan pieces have been selected in the display menu, it will turn from light to dark blue, see Fig. 6.23. More than one variant can be created in a model, and the piece can be included in each variant as required.

Chapter 8

The upper tool bar in Lectra Modaris pattern cutting software

Abstract: This chapter explains the Modaris upper tool bar, its drop-down menus and keyboard shortcuts. Each menu is explained in turn and referenced to Chapters 6 and 7 where appropriate. Functions identified as advanced because of their complexity or disused due to upgraded versions of the software will not be covered.

Key words: upper tool bar menus, prompt box, dialogue box, parameters, Lectra Modaris.

8.1 Upper tool bar

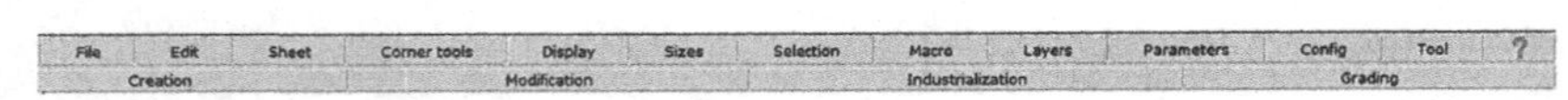

8.1 Upper tool bar menu bar

Across the top of the screen is a tool bar with 13 separate drop-down menus. Below are four macro menus which will not be covered here. Once you are familiar with the main menus and functions, the macro menus will be a natural progression. Here, only the most frequently used and need-to-know commands in general pattern making are covered.

Where there is a keyboard shortcut available, this is shown in brackets immediately after the function title. Be aware that many of these shortcuts are case sensitive, making the keyboard shortcut less favourable. However, where it is considered the more convenient option, usually because both the upper and lower case letter work, it is displayed in **bold**.

Items that are no longer used due to the upgrading of Modaris, are more complex and consequently for advanced use or are to do with grading will not be covered.

8.1.1 *Prompt box*

Commands that are irreversible will call up a **prompt box** to check that you really want to execute the specific command (Fig. 8.2). **Saving** and therefore overwriting a previously saved file, or **deleting** work, are the two most common, but there are others which will alert you to an unforeseen consequence of the intended action. Read the **prompt box** and select **Continue** or **Cancel/Abort** as required.

8.1.2 *Dialogue box*

Many of the commands in this section will open a **dialogue box** in which to enter a code or measurement value (Fig. 8.3). The format will vary according to the function requirements. When using a dialogue box, remember the following:

- click into the field and enter the required figure,
- use the **arrow keys** to move between one field and another,
- use **Enter** to complete the command.

See Chapter 6, Section 6.1.1 for a more detailed explanation.

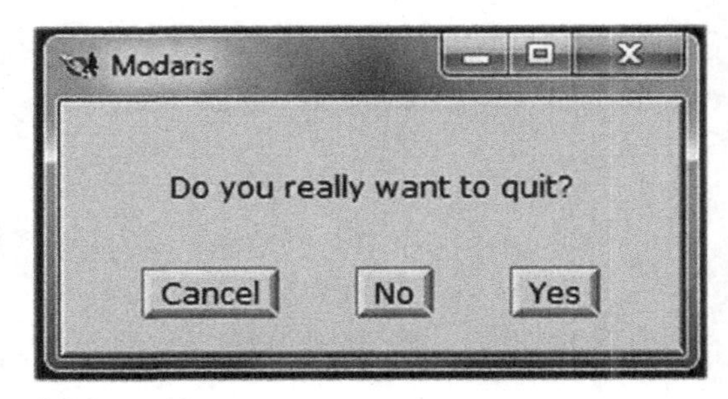

8.2 Prompt box

8.3 Dialogue box

8.2 File menu

Please refer to Fig. 8.4 for this section.

8.2.1 New (Ctrl + N)

New (**Ctrl** + **N**) is selected to create a new file. A **dialogue box** will open in which to enter the new name or number. Press **Enter** to complete the function.

It is important to note that the first time a **new** file is saved, no **prompt box** should appear. If a **prompt box** does appear, it means that the file name you have chosen already exists and you will overwrite and lose the original file, if you continue. Select **Cancel/Abort** and start again with a new file name.

8.2.2 Open model (Ctrl + O)

Open model (**Ctrl** + **O**) is used to open an existing file.

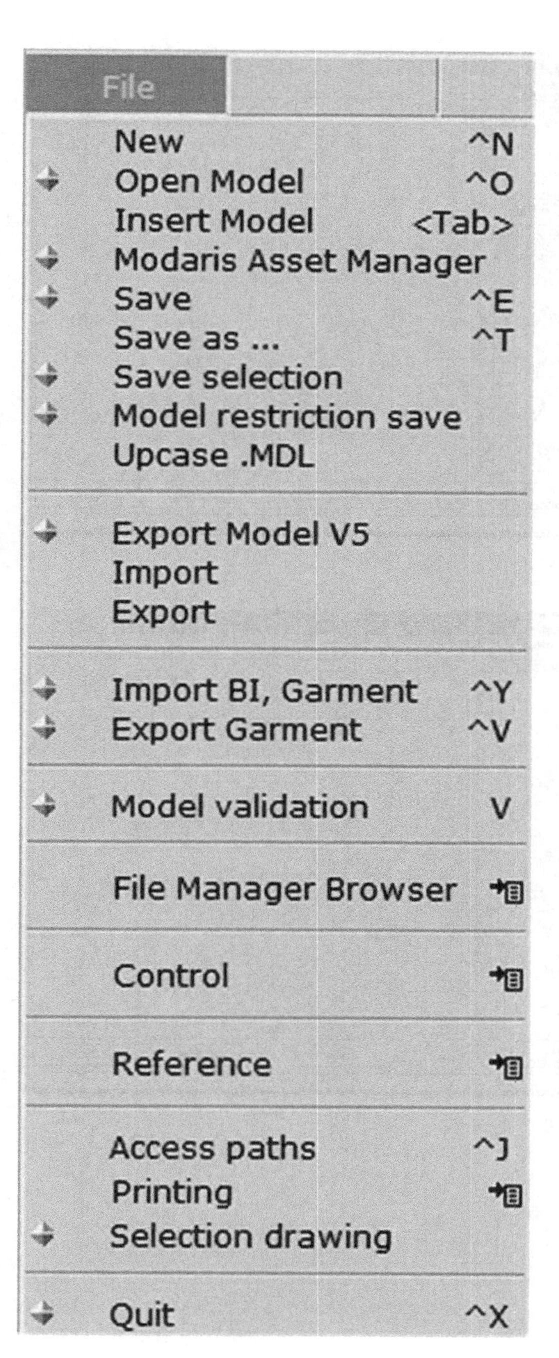

8.4 File menu

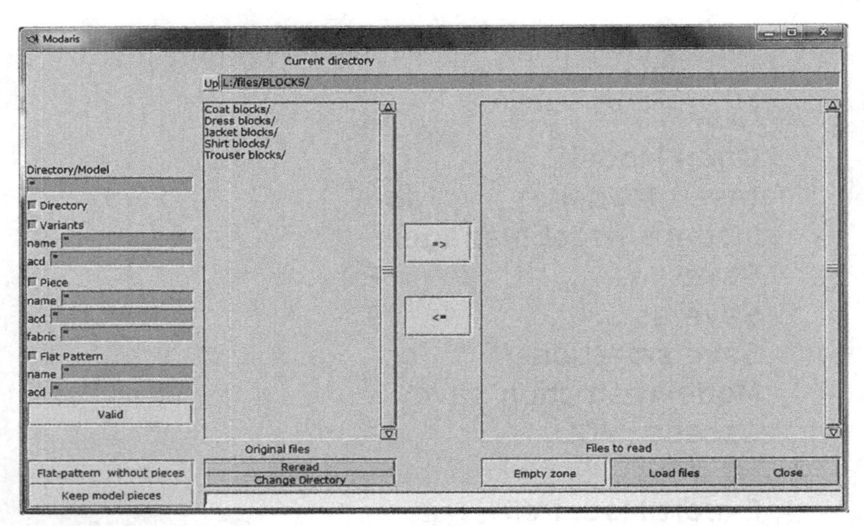

8.5 Model insert opening window

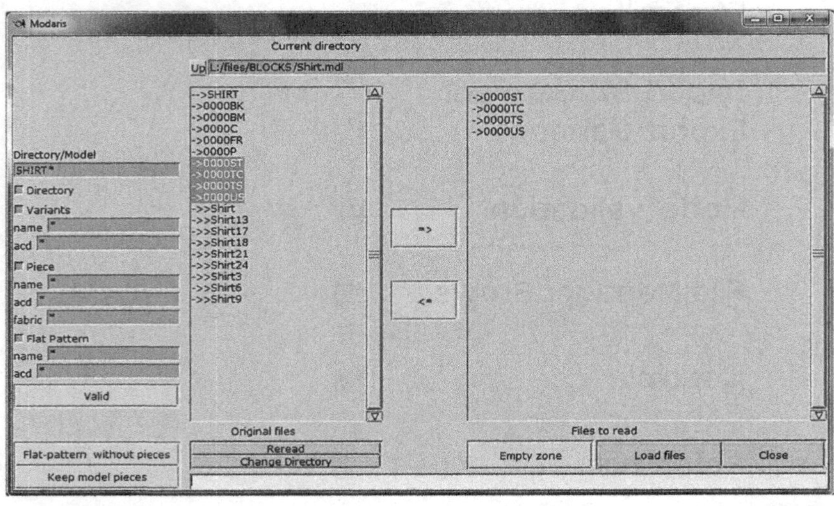

8.6 Model insert window showing selected items

8.2.3 *Insert model (tab key↹)*

Insert model (**tab key↹**) is used to insert existing pieces or a complete model into the open model. When **tab↹** is activated, a new window will open in which to make your selection. It is here that disciplined naming and filing will pay off (Fig. 8.5). Remember the following general rules:

- use **Up** to access other file libraries,
- use **Directory/Model** to enter a specific file in the selected library,
- use **Valid** to search for the model,
- use **Reread** to go back one stage.

To use this function, **double click** on the folder name to access the file. Double click again if there is another layer. Once the required file is displayed, the whole model can be imported to the open file. Alternatively, double click on the file name to select a single piece or selection of pieces (Fig. 8.6).

Highlight the piece required. Hold **Shift** for multiple selection.

Use the central **blue arrow** button to transfer the selected pieces to the right-hand field. This action can be repeated as necessary.

Use **Load files** to add the pieces to the open model.

Use **Close** to finish and close the window.

The pieces will then be displayed in the current model and can be renumbered and used as required.

8.2.4 *Save (Ctrl + E)*

Save (**Ctrl** + **E**) will **save** the open model to the default library.

8.2.5 *Save as (Ctrl + T)*

Save as (**Ctrl** + **T**) allows the **save** to an alternative library.

8.2.6 Access paths

Access paths is used to define or change the **save** and **retrieve** pathways. Understanding how to manage and change **access paths** is key to managing your files, and will reduce frustration when certain functions appear not to work.

Figure 8.7 shows the **design** folder as the default setting for accessing and saving all files. This is a practical setting when most work is saved to **design** as the current working file. However, this can be altered to your requirements.

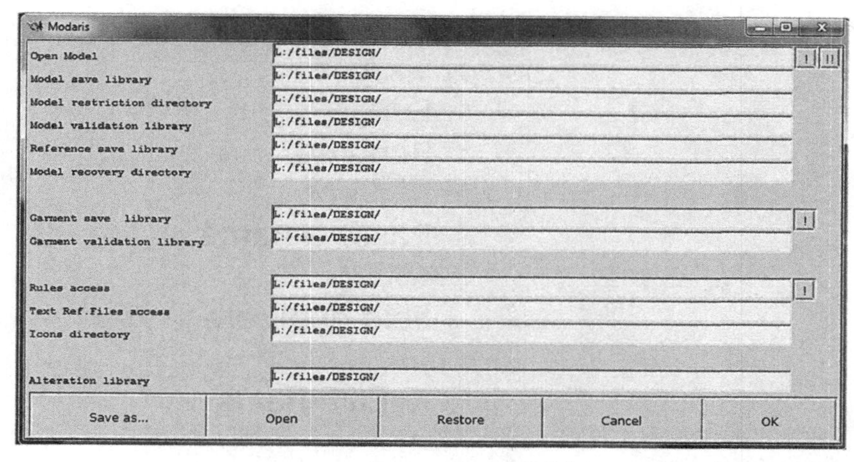

8.7 Access paths selection window

To change a pathway, click in the top field **open model**, to open the **browse for folder** window (Fig. 8.8) and select as required. The browser field will close and take you back to the **access paths** window. Please note:

- use ! to auto fill the top fields,
- use !! to auto fill all the fields.

8.8 Access paths browser for folder window

Please note the following about the folder examples shown here, suitable for single client use:

- **archive** is for keeping patterns more than four seasons old,
- **blocks** is for all the basic pattern blocks,
- **design** is the active folder for current work,
- **design 1st cut** is for holding a copy of the original pattern created in **design** while and until any amendments are carried out and approved. It is surprising how often the original cut is needed, so always keep a copy,
- **designarc01/designarc02/designarc03** are recent past season's patterns.

Lectra recommend limiting the number of files in a folder to maintain an efficient working speed. To do this it is necessary to update the file libraries from time to time. Using **Windows Explorer**, the files can be easily transferred at the end of the season or year to maintain **design** as the current working folder. One way to do this is to transfer all the **designarc03** files to archive then:

- delete the empty **designarc03** folder,
- rename **designarc02 as 03**,
- rename **designarc01 as 02**,
- rename **design as designarc01**,
- create a new **design** folder.

A large pattern room making patterns for many different customers may decide to organise a folder for each client. Creating a system to which all users comply is essential for the good management of the pattern room. This need not be complicated. However, once patterns are graded and markers created in Diamino, it becomes imperative that a system is in place and followed.

8.2.7 Quit ☒ (Ctrl + X)

Quit ☒ (**Ctrl** + **X**) will close the current file. So that you cannot close a model accidentally, a **prompt box** will appear for confirmation of the action. Read the **prompt box** and act accordingly.

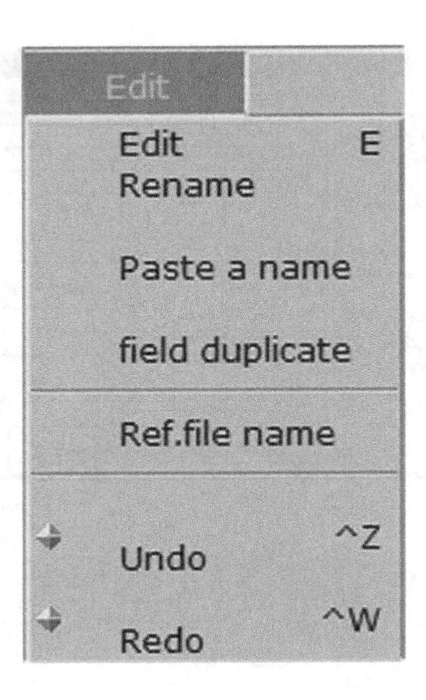

8.9 Edit menu

8.3 Edit menu

Please refer to Fig. 8.9 for this section.

8.3.1 Edit (E)

Edit (**E**) enables writing and editing text in the **title blocks** and onto the **special axis**.

8.3.2 Rename

Rename is for the renaming of pieces within a model. This function allows the retention of all of the pattern piece **codes** and **comments** information as well as the **variant**, while copying a model to create a new one. This is useful when using blocks that now need a working file code or when making minor style alterations to an existing model which then becomes a new style.

To use **Rename**, open the model to be copied. Use **Save-as** and assign a new number. Place the cursor over the **model identification page** and the new number will be displayed in the interactive bar just above the working screen. However, the pieces within the model will still have the previous name and need to be amended. Now use **Ctrl** + **A/a** to select all sheets, and then activate **Rename.** Use the **dialogue box** to define the current file name (string) and the new name. Press **Enter** to complete the action. **Save** (see Fig. 8.10).

8.3.3 Undo (Ctrl + Z/z)

Undo (**Ctrl** + **Z/z**) will undo the last command and up to 20 steps.

8.3.4 Redo (Ctrl + W/w)

Redo (**Ctrl** + **W/w**) will redo the last command.

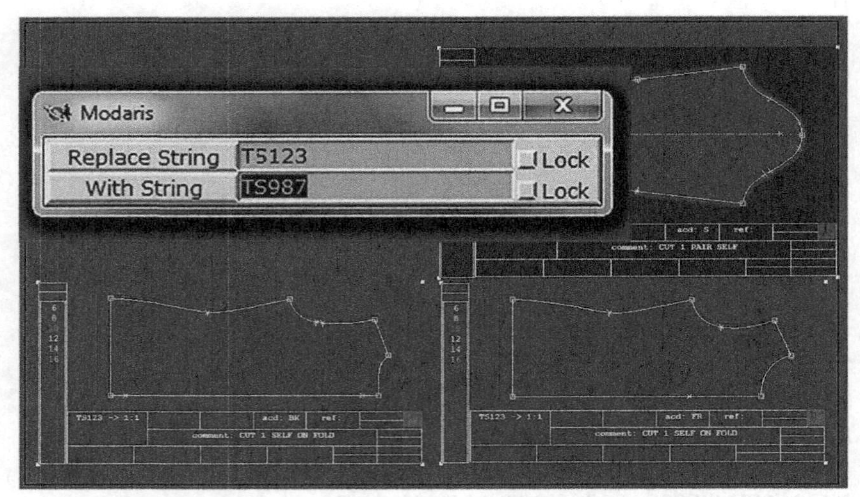

8.10 Rename dialogue box

8.4 Sheet menu

Please refer to Fig. 8.11 for this section.

8.4.1 New sheet (N) to create a new sheet

New sheet (**N**) is used to create a new sheet. Note that general typing may activate this function unintentionally. Use **Sheet – Delete** to remove any unwanted new sheet.

8.4.2 Copy (Ctrl + C) to make a copy of a sheet

Copy (**Ctrl** + **C**) is used to make a copy of a sheet. Click on the sheet to be copied. Right click to complete. Be aware that the new sheet is not always immediately visible, so press **J/j** to bring it into view. A piece already selected for the **variant** will be dark blue (provided **orphan pieces** is ticked in the **Display** menu), but the copy will be light blue, see Section 8.6.5. When making pattern amendments, first make a copy to work on and, when ready to replace the original piece, use **F4 – Exchange data**.

8.4.3 Delete (z) to delete a sheet

Delete (**z**) to delete a sheet. Hold **Shift** and the flat pattern layer will be deleted at the same time. **Ctrl** + **Z** will undo the action but you will need to take a copy of the retrieved sheet to ensure its stability.

8.4.4 Sheet Sel. to select a sheet

Sheet Sel. is used to select a sheet or number of sheets, most commonly to view in a new window. This is useful when there are too many pattern pieces on the desktop. Activate **Sheet Sel.** and the cursor will become a pointing finger icon. Click on the centre of each piece that you want to work with and each sheet will display small white squares at each corner. To de-select a piece, left click on it while **Sheet Sel.** is active.

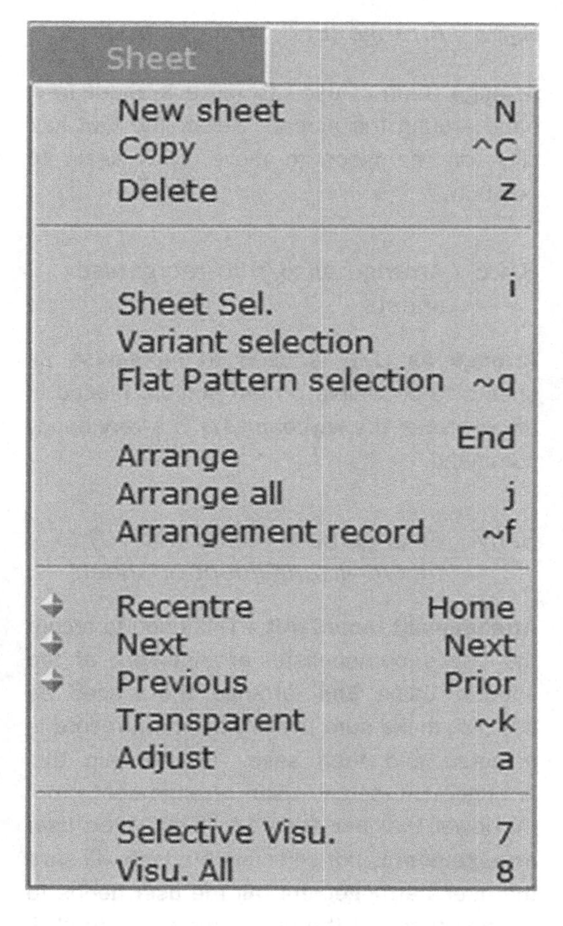

8.11 Sheet menu

Move the cursor away and outside of the black line that marks the edge of the working page and **right click**. The finger icon will turn to an arrow icon (Fig. 8.12). Press **7** on the main keyboard, and a **dialogue box** will appear in the top left-hand corner. Ignore this and click in the centre of any one of the selected sheets, and a new window will now open displaying only the selected sheet or sheets. Use **J/j** to **re-arrange all** while in this window. Use **8** to return to the main window.

8.12 Sheet select cursor icons

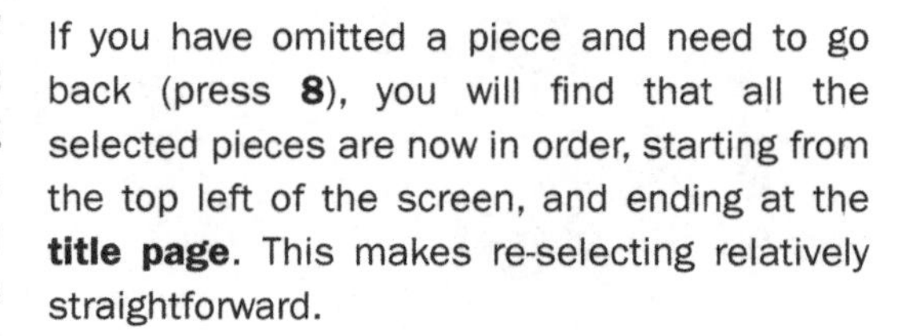

If you have omitted a piece and need to go back (press **8**), you will find that all the selected pieces are now in order, starting from the top left of the screen, and ending at the **title page**. This makes re-selecting relatively straightforward.

8.4.5 Arrange (End) to move a sheet

Arrange (**End**) is used to move a sheet free-hand around the screen. Select the **End** key, click on the piece to move, click again to position.

8.4.6 Arrange all (J/j) to reorganise sheets

Arrange all (**J/j**) is used to reorganise all sheets to be visible on the screen. Placed in the centre of the keyboard, **J/j** is a very handy command.

8.4.7 Arrangement record (Alt + f) to record arrangement of sheets

Arrangement record (**Alt + f**) is used to record the user's personalised arrangement of the sheets. Using **End**, arrange the pieces as desired, make sure the arrangement record is selected and then **save**. To maintain this arrangement select **User arrangement** from the lower tool bar, then **J/j**. Even when **user arrangement** is not selected, Modaris will save the user's arrangement. All the user needs to do next time she/he opens the file is to click **user arrangement** to view her/his previous arrangement.

8.4.8 Recentre (home) to centre sheets

Recentre (**home**) will bring the selected sheet to fill the screen.

8.4.9 Next (page down) to scroll through sheets

Next (**page down**). This is a handy way to scroll through all of the sheets when using the same function, i.e. adding the grain line, filling in **text boxes** or final pattern checking.

8.4.10 Previous (page up) to scroll through sheets

Previous (**page up**) is another way to scroll through sheets, as above.

8.4.11 Adjust (a) to adjust or resize a sheet

Adjust (**a**) to adjust or resize a sheet back to its smallest size. Use after the **divorce** function.

8.4.12 Selective visual (7) to isolate a chosen sheet

Selective visual (**7**) is used to isolate a chosen sheet or sheets to a new window. See 8.4.4 **Sheet sel.** above.

8.4.13 Visualise all (8) to return to the main window

Visualise all (**8**) is used to return to the main window.

8.5 Corner tools menu

This is an advanced menu for customising corner options (Fig. 8.13). Although corners can be selected using this menu, it is more convenient to use **F4 – Change corner**, or the **corner tools** selector on the lower tool bar. The Lectra library of corners is suitable for most applications but there is just one, **mitered**, which needs customising.

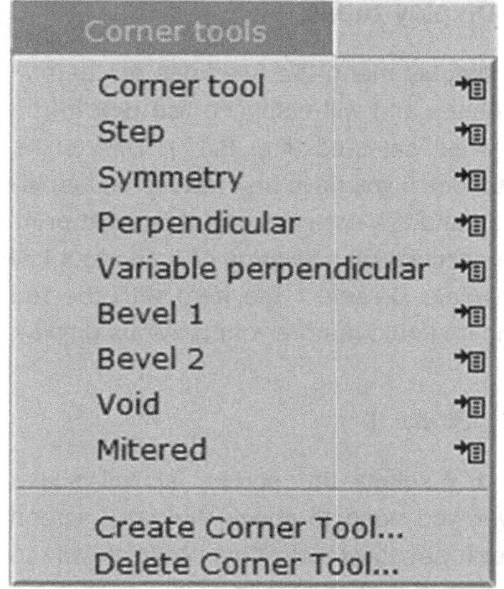

8.13 Corner tools menu

8.5.1 The standard mitre corner

The standard **mitre** corner, introduced with Modaris Version 6, will trim the excess diagonally across the corner, but without allowing any seam allowance (illustrated by the red line in Fig. 8.14A).

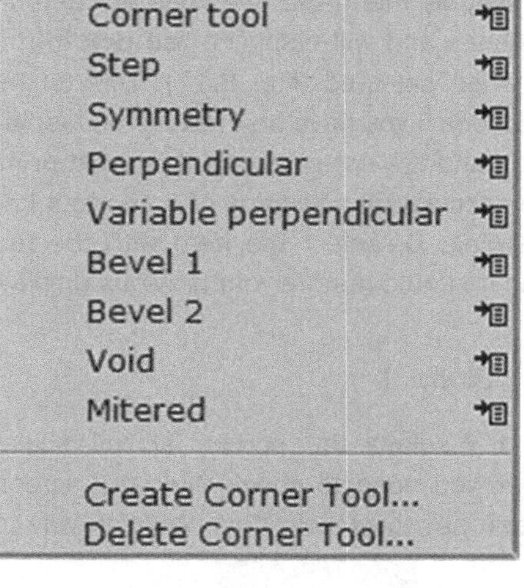

8.14 Before (A) and after (B) a mitre corner amendment

To add seam allowances, it is necessary to customise the corner by adding seam allowance values as shown in Fig. 8.14B.

Under the **Corners tool menu** select **Create corner tool** and a dialogue box will open. Place the cursor in the **Generic corner** field and use **tab**⇆ to access the corner menu options. Select **Mitered** and press **Enter** (see Fig. 8.15).

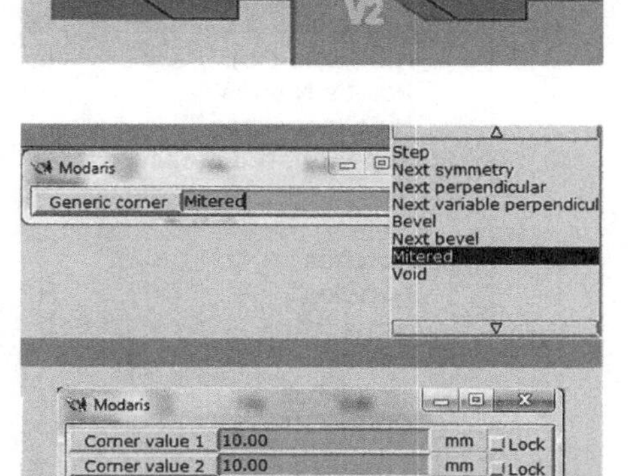

8.15 Mitre corner dialogue boxes

A further window will open, into which **corner values** (seam allowances measurements) relating to the existing lines are entered. There are three value options that can be added, but for this example only V1 and V2 are needed.

Press **Enter** and the window will close and the new corner filed in the **Corner tools – Mitered** menu. Select the new corner and click on the corner to be amended (see Fig. 8.16).

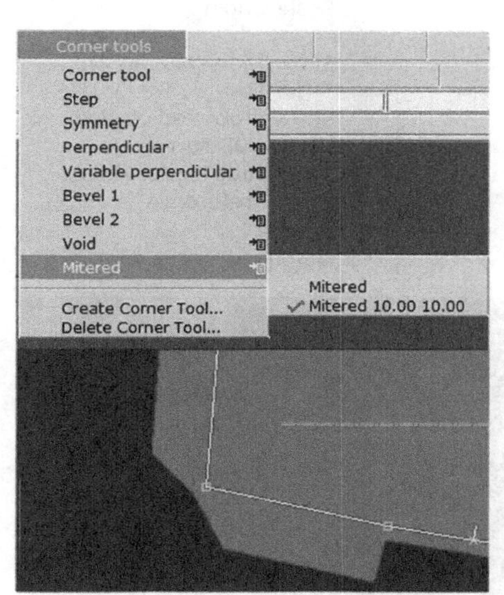

8.16 Customised mitred corner

The same general principles are applicable for customising all corner types, and it is a case of trial and error with the value options until the required shape is attained. Note that any customised corner is filed in the **Corner tools menu** in a sub-section accessed by scrolling down to the corner type and then moving the **cursor** to the right.

8.6 Display menu

In the **Display** menu the functions act as on/off switches and will display a **red tick** in the menu when selected (Fig. 8.17). This is the place to look if you have an unexpected display variation and are not sure why. It is most probably the accidental activation of a shortcut key while typing. De-select the item with the red tick and it should restore your previous display.

8.6.1 Scale 1

Scale 1 displays the screen at full scale. However, you need to check that your screen has been set to achieve this. A quick way to verify this is to draw a 20 cm square (**F2 – Rectangle**). Select **Scale 1** and then measure the square externally with a ruler. It should measure 20 cm.

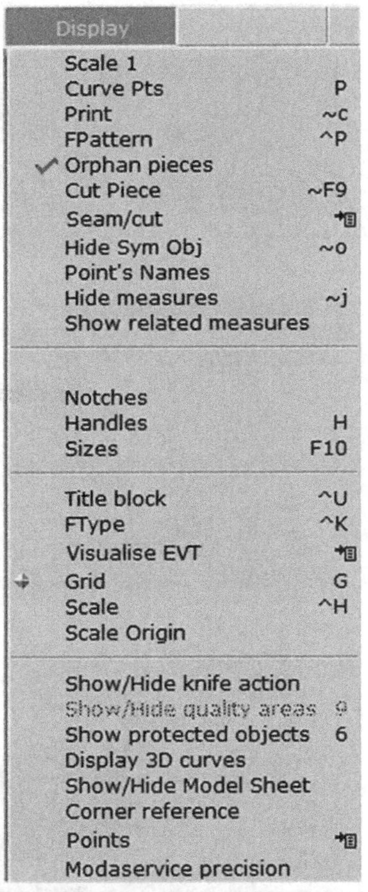

8.17 Display menu

8.6.2 Curve points (P)

Curve points (**P**) is also available on the **lower tool bar.** This is an on/off control used to display the **curve points**, displayed as red points.

8.6.3 Print (Alt + c)

Print (**Alt + c**) records the position of the active lines and points so that when any one of them is moved, a purple shadow line remains. This is a very useful function with many applications. Before making any amendments to a pattern piece, activate **Print**. This will allow you to see exactly how and what you are moving. To compare pattern pieces, use **Marry**, then turn on the **Print** function and **Divorce** (or **Ctrl + Z**) the piece. An outline of the piece will remain. Be aware that, once the print is turned off, the shadow line will disappear and cannot be retrieved.

8.6.4 FPattern (Ctrl + P)

FPattern (**Ctrl + P**), also available on the **lower tool bar**, is an on/off control to display the flat pattern layer, in other words the shape and history (root) from which the current piece has been extracted.

Occasionally a sheet will not **adjust** (**a**) back to its smallest, or there is something invisible hindering an action.

Figure 8.18 shows a pocket piece that has been extracted from a front. The armhole and side seam have remained on the flat pattern, but are only visible when **FPattern** is activated. Often this has no effect on the actual pattern piece but, occasionally, it may prevent a function. Turn on **FPattern** and delete any unwanted lines or points, which may be causing the problem.

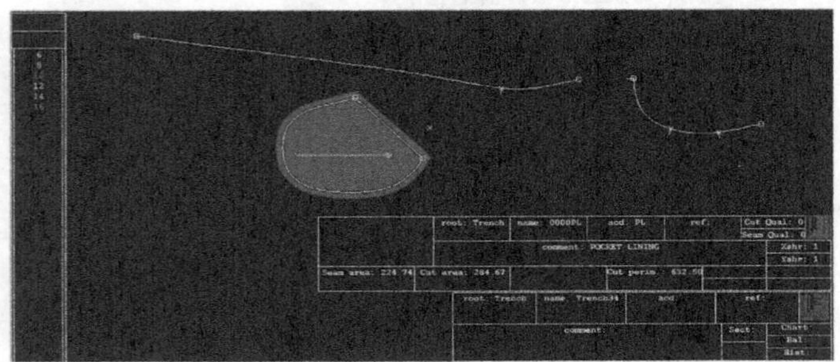

8.18 FPattern layer

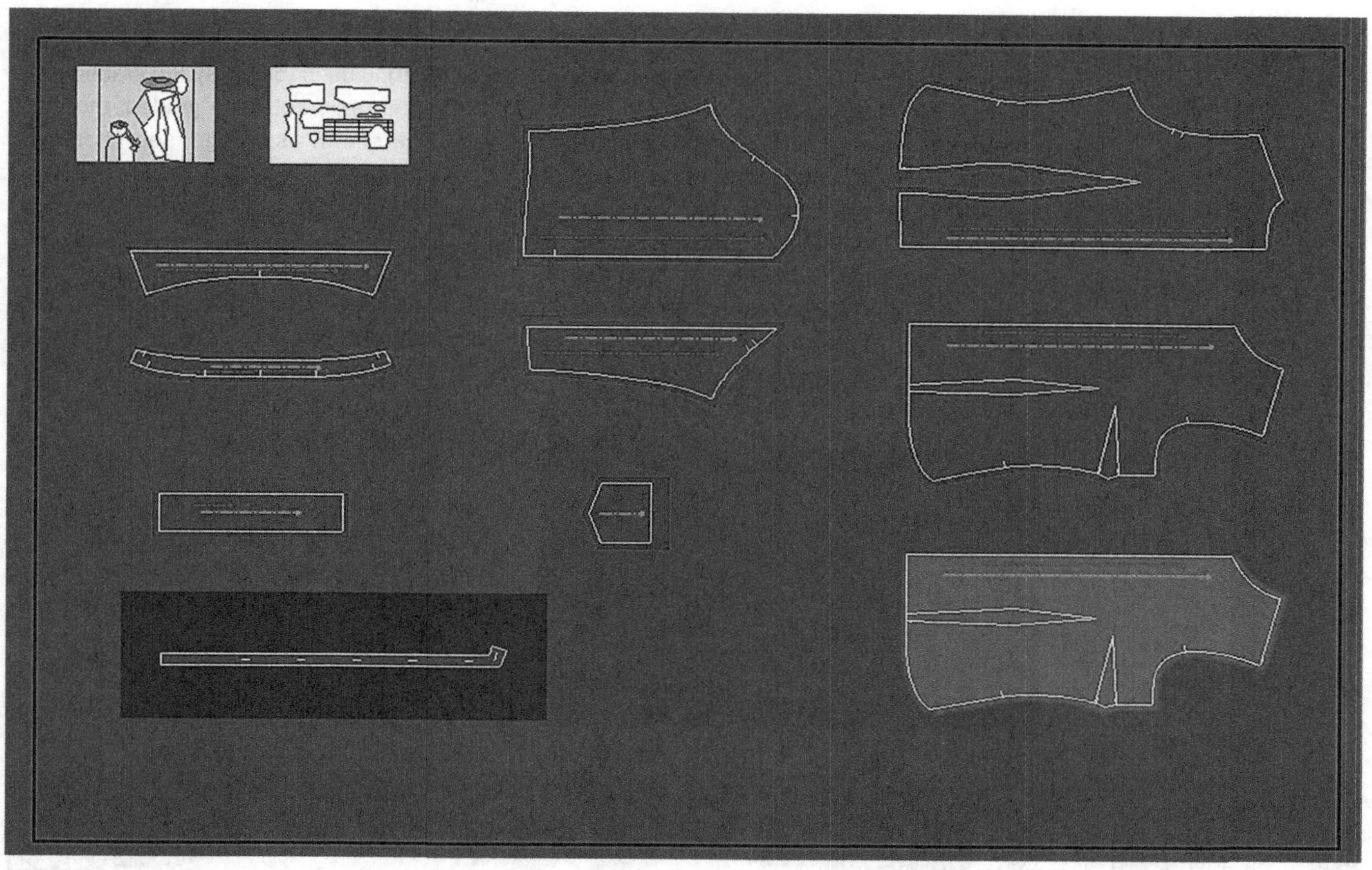

8.19 Orphan pieces

8.6.5 *Orphan pieces*

Orphan pieces is an on/off control to identify sheets that have been selected for the **variant**. The selected sheets turn dark blue (Fig. 8.19).

8.6.6 *Cut piece (Alt + F9)*

Cut piece (**Alt** + **F9**), also available on the **lower tool bar**, displays the outer or cut edge of the pattern piece. These lines are displayed in red. Turn off again to continue working, see Fig. 8.34 in Section 8.10.1.

8.6.7 *Hide measures (Alt + j)*

Hide measures is used to hide or display on screen the pattern measurements taken with the **dynamic measurements.** The use of **dynamic measurement** menu is not covered here (see Fig. 8.20).

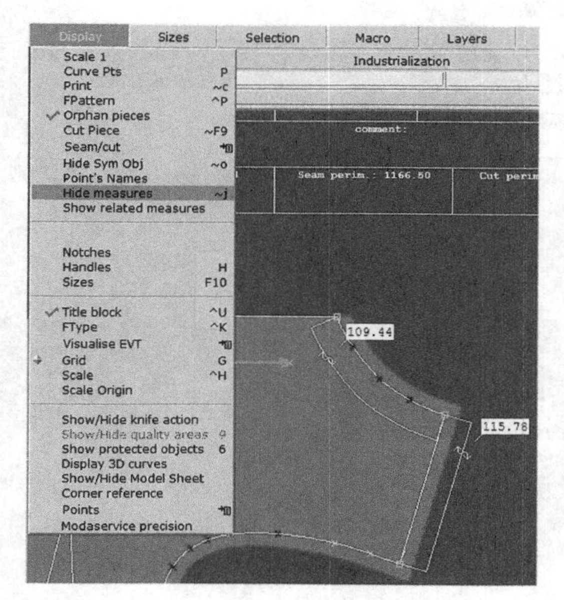

8.20 Hide measures

8.6.8 Handles (H)

Handles displays the handles on **Bezier curves**, allowing for fine adjustment of the curve without actually moving any of the points. The handles show as short tangents at the point position. Click onto one of the handles and rotate until the required curve shape is obtained. Note that this only works on Bezier curves, not normal curve points (see Fig. 8.21).

8.21 *Bezier curve handles*

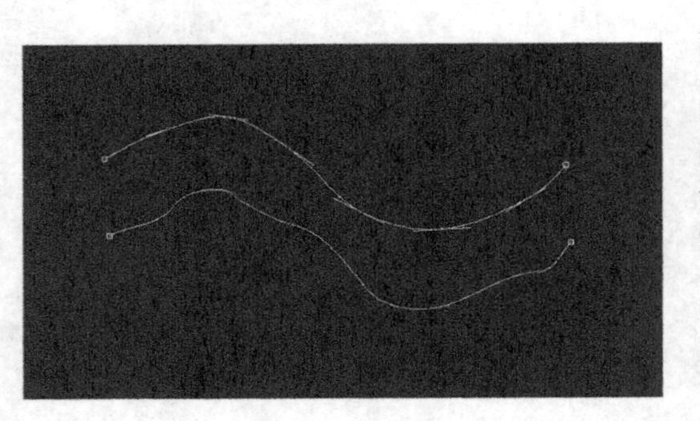

8.6.9 Sizes (F10)

Sizes (F10) is used to display the graded nest. If the pattern has not been graded, the outer line will be shown as orange, see Chapter 7, Fig. 7.17, front A.

8.6.10 Title block (Ctrl + U)

Title block (**Ctrl** + **U**) will reveal the **title blocks** containing text around the sheet. See Chapter 3, Fig. 3.1.

8.6.11 FType (Ctrl + K)

FType (**Ctrl** + **K**) displays the sheets in colour, depending on fabric type indicated in the variant. Note that the piece not included in the **variant** is now dark blue (see Fig. 8.22).

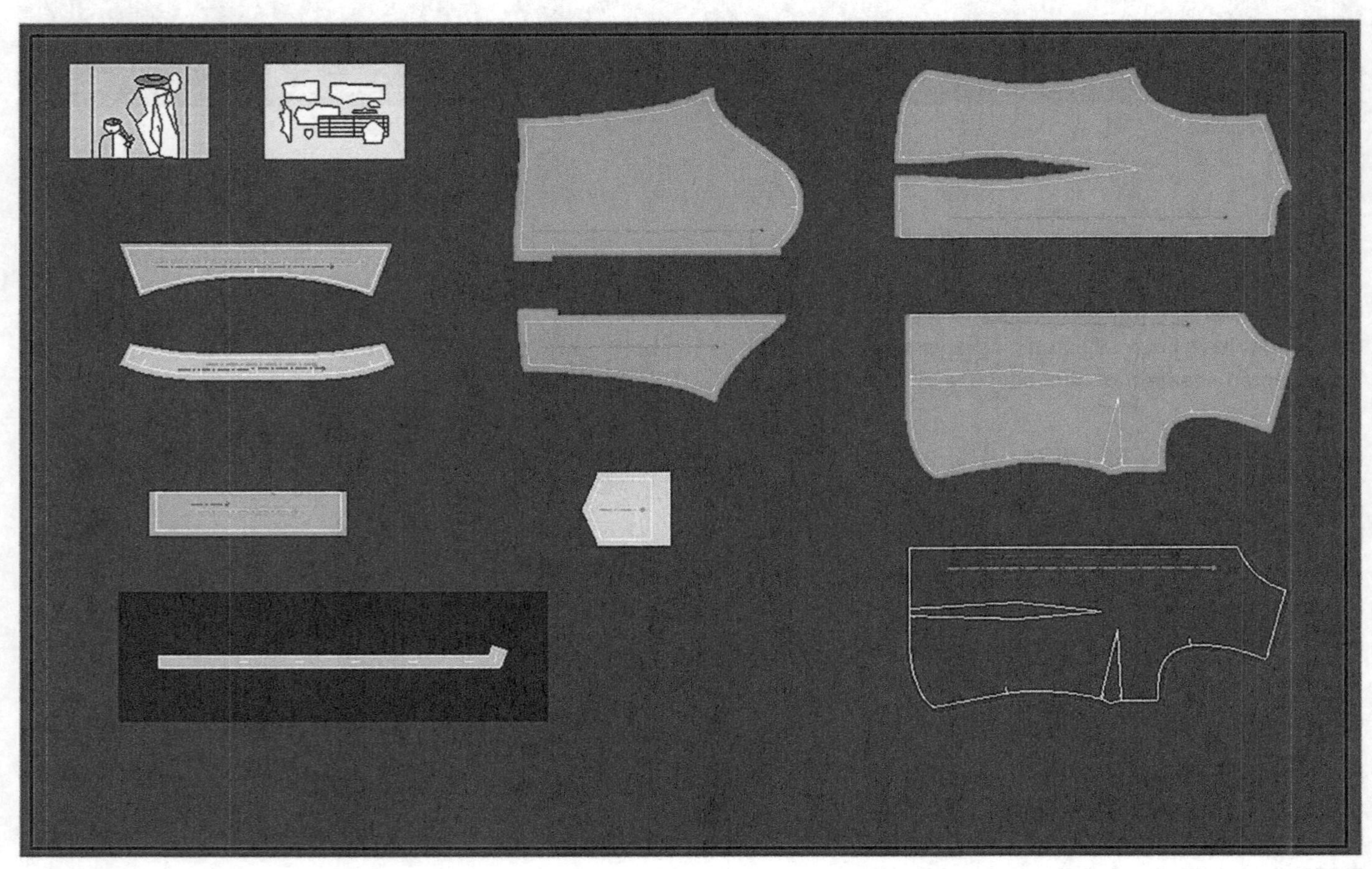

8.22 *FType*

8.6.12 Grid (G)

Grid (**G**) is an on/off control to display a grid of dots on the screen. Please be aware that shortcut **G** may be activated unintentionally (see Fig. 8.23).

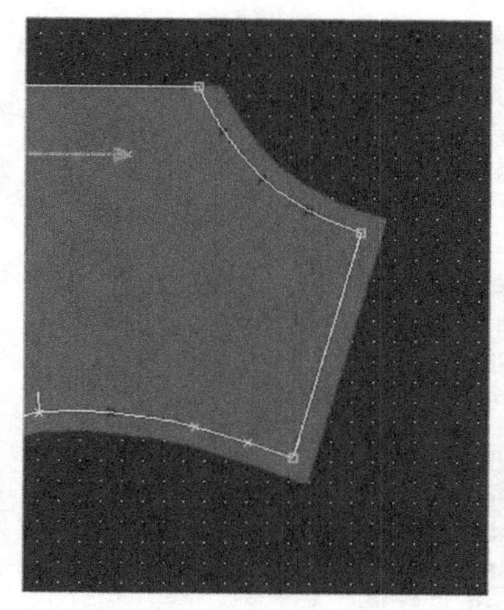

8.23 *Grid*

8.6.13 Scale (← backspace key or Ctrl + H)

Scale (**← backspace key**) is an on/off control to display a **scale** ruler on the **x**- and **y**-axes, along the border of the **title boxes.** This can easily be activated unintentionally, see Fig. 8.24 and Fig. 8.25.

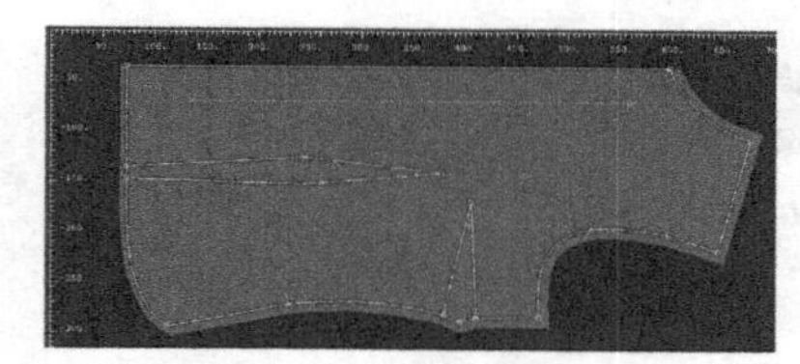

8.24 *Scale*

8.6.14 Scale origin

Scale origin allows the positioning of the scale as required. Activate **scale** (use **backspace**) first, then select **Scale origin** and click on the required point of intersection of the *x*- and *y*-axes.

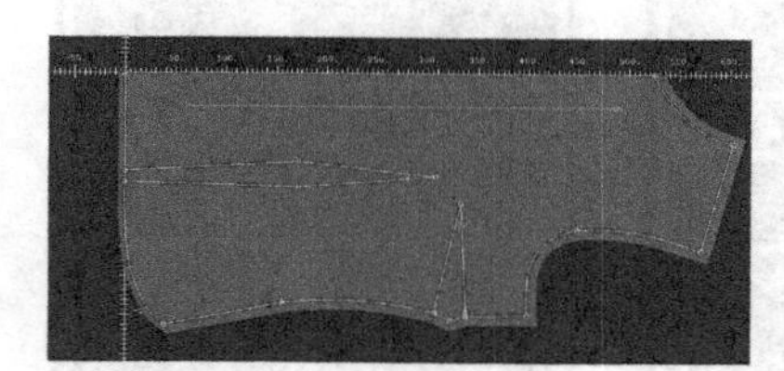

8.25 *Scale origin*

8.7 Sizes menu

This menu is associated with **F7 – Evolution system functions** used for naming and defining size ranges (Fig. 8.26). Refer to Chapter 11, Section 11.6 Size correspondence.

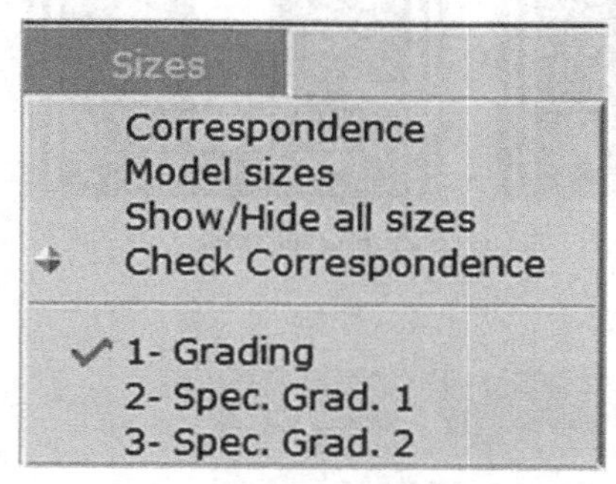

8.26 *Sizes menu*

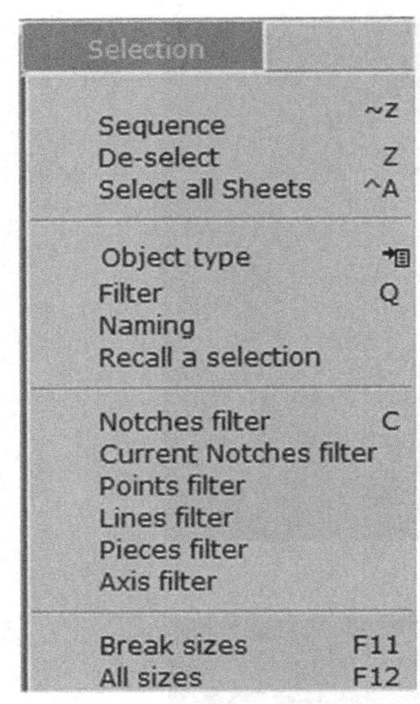

8.27 Selection menu

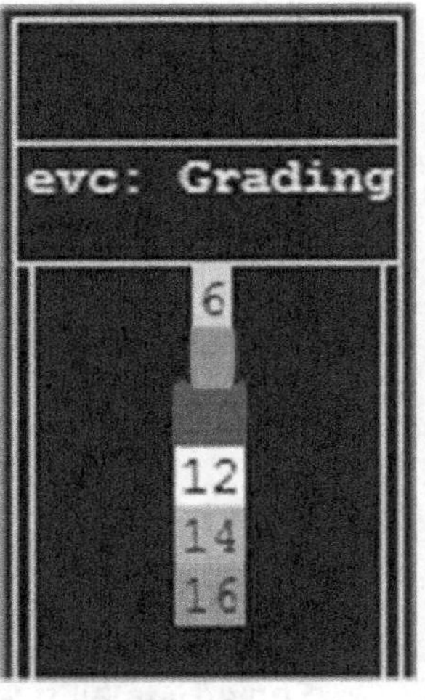

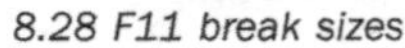

8.28 F11 break sizes *8.29 F12 all sizes*

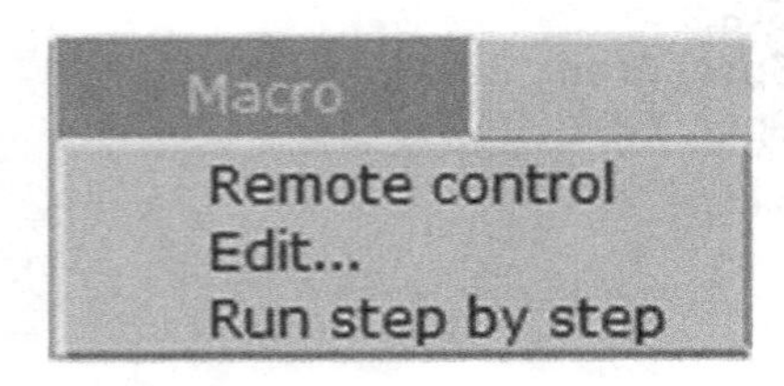

8.30 Macro menu

8.8 Selection menu

Please refer to Fig. 8.27.

8.8.1 De-select (Z)

De-select (Z) – Used to remove the selection function. More convenient is the **selection** field at the top of the **functions menu** column. The **Esc** key can also be used to exit a function.

8.8.2 Select all sheets (Ctrl + a/A)

Select all sheets (**Ctrl** + **a/A**) is used to select all the sheets, identified by the white squares displayed at each corner. See **Edit – Rename** (Section 8.3.2) for how to use when copying and re-numbering a whole pattern.

8.8.3 Break sizes (F11)

Break sizes (**F11**) is the default setting and generally does not need selecting. When using the graded nest display (**F9**) or taking measurements on screen (measurement options accessed in **F8**), only the break sizes will be displayed or highlighted, see Fig. 8.28 and Fig. 8.29.

8.8.4 All sizes (F12)

All sizes (**F12**) is selected immediately before using another command so that the whole size range can be viewed in that particular command. All of the sizes listed on the left in the text box will become highlighted. The most common uses are for the **F8 – Spreadsheet**, when taking measurements, or to display the complete graded nest when using **F9**.

8.9 Macro menu

This menu is associated with the macro tool bar positioned immediately below the upper tool bar and is for the advanced user (Fig. 8.30).

8.10 Layers menu

This menu is for managing the different layers of a pattern (Fig. 8.31) widely used by Modaris Expert users, and will only briefly be covered here. Modaris Expert is an advanced module of Modaris.

8.10.1 Layer types

A sheet is made up of a number of layers superimposed on top of each other. The user can, if desired, add more layers. In some specific garment types this is a useful application. However, to begin with, the default layers are sufficient.

The **construction layer** (1) displays only a colour block shape hiding the working lines. This layer may be displayed accidentally by pressing number **1** on the keyboard. Press **2** to return to the base layer (see Fig. 8.32).

The **base layer** (2) is the normal working layer recognised as a grey sheet background containing the blue pattern shape with a white line defining the digitised edge. This is the layer to add grain and comments axis to in **F4** (see Fig. 8.33).

Cut piece on the lower tool bar will show the actual outline of the piece as it will be plotted. This displays the cut edge in red, showing all of the corners and notches as they will be plotted out, including the comments from any axis. Turn this function off to return working.

Display seam/cut lines from the layers menu displays almost the same thing but with a slightly different configuration (see Fig. 8.34).

8.10.2 Layers menu options

Create, Layer rename do what they say in the same way as any computer program.

Insert and **Erase** also do what they say.

Hidden objects (**Alt + 0**) will display any points, lines or marks not visible on the working layer. These hidden objects may prevent a function working properly, so just delete as necessary (see Fig. 8.35).

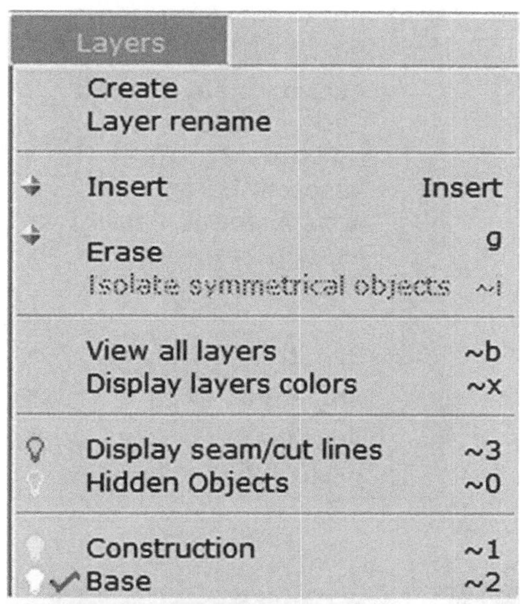

8.31 Layers menu

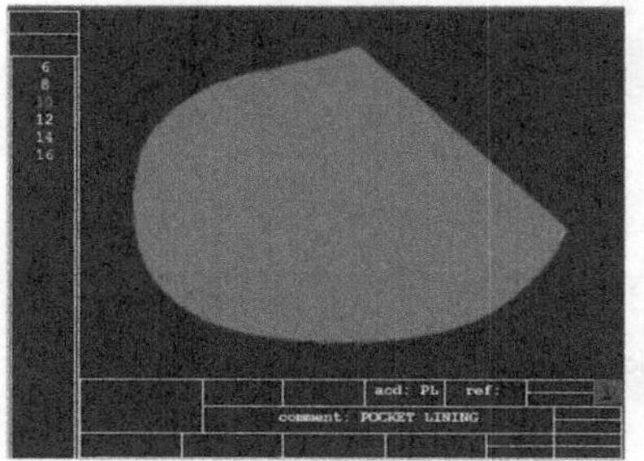

8.32 Construction layer

8.33 Base layer

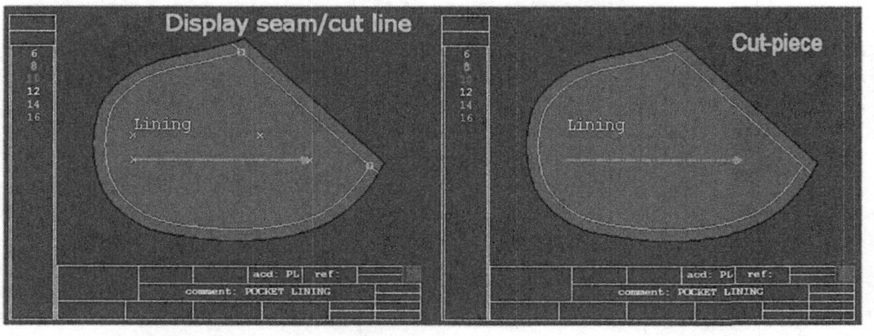

8.34 Display seam/cut line and cut piece

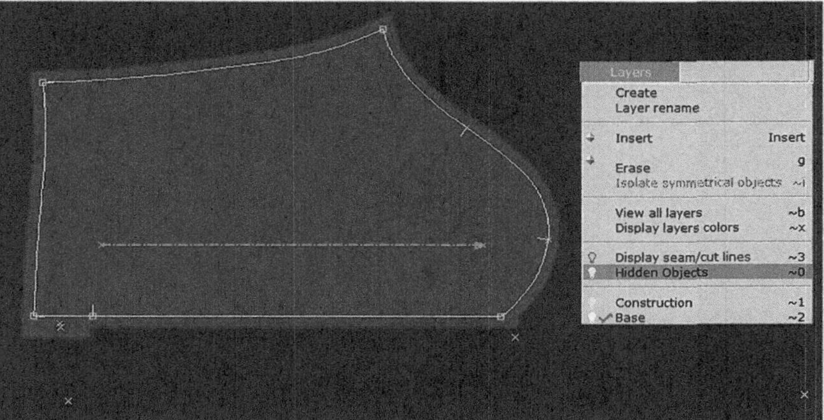

8.35 Hidden objects

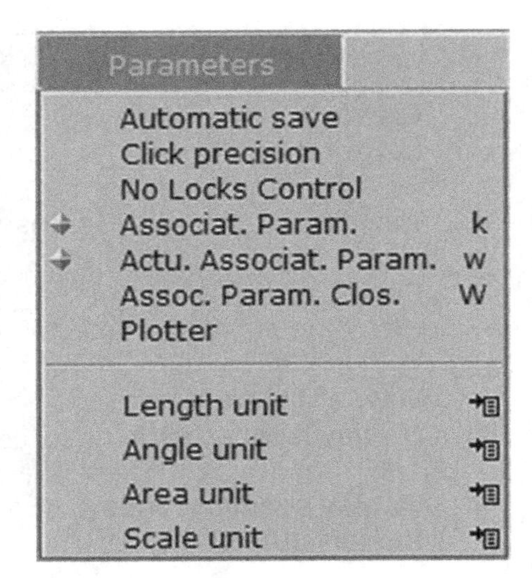

8.36 Parameters menu

8.37 Automatic save dialogue box

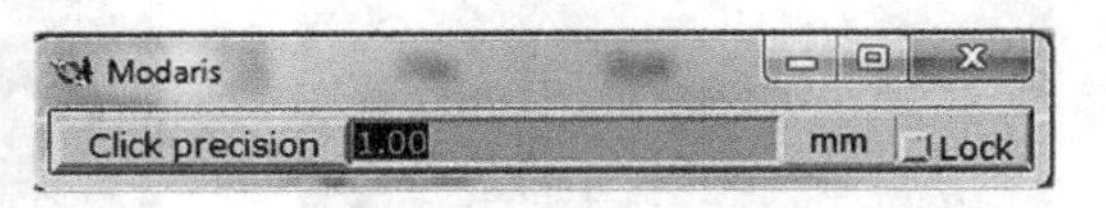

8.38 Click precision dialogue box

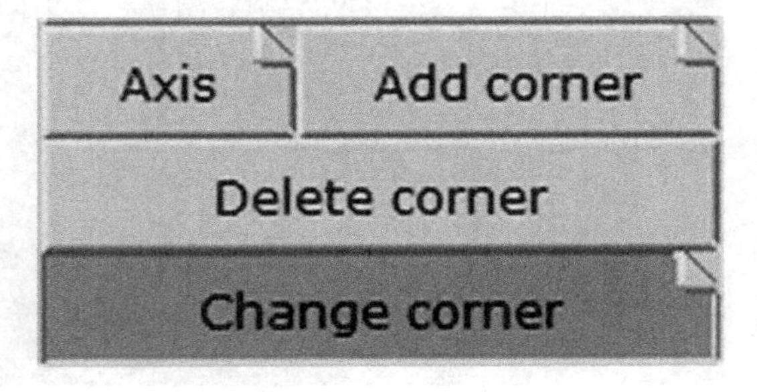

8.39 Associated parameters access via corner of function field

8.11 Parameters menu

This is a seldom used but important menu where the controls of the overall working environment structure are defined (Fig. 8.36). Once set to your requirements, you may never need it again. However, if you make patterns for a number of different clients, you may have varying parameter requirements. As an example, one client may work in inches and another in centimetres.

8.11.1 Automatic save

Automatic save saves work at regular intervals as defined by the user. Use the dialogue box to specify your requirements, defined in seconds (see Fig. 8.37).

8.11.2 Click precision

Click precision is used to define the radius or distance from the point in which the association will be made. In other words, it defines how close you need to click beside a point for it to link to that point (Fig. 8.38).

8.11.3 No locks control

No locks control is an **on/off** button for use when more than one user is working from the same server. When **on** it will inform the user when a model has already been opened by someone else.

8.11.4 Associate param (k) and Actu Associat Param (w)

Associate param (**k**) and **Actu Associat Param** (**w**) are an alternative choice to the turn down corner of the functions fields to access the further options (**parameters**) available in that specific function. For example, the **change corner** has further choices (see Fig. 8.39).

Other functions are not so obvious. To discover if there are further choices within a function, first select the function and then select **Associate param.** A **dialogue box** will open with the further options. If there are no further parameters, the window will say **no associated parameters**.

Note that when pressing **J/j**, it is possible to miss the key and select **k** instead. This is the **associate parameters** shortcut. This will open the **dialogue box** unexpectedly. Click **ok** and it will close. See Chapter 6, Section 6.1.2 for more details.

Assoc param close will close the **dialogue box**.

Length unit is where to make the choice not only between metric and imperial but also of the size of the base unit (see Fig. 8.40).

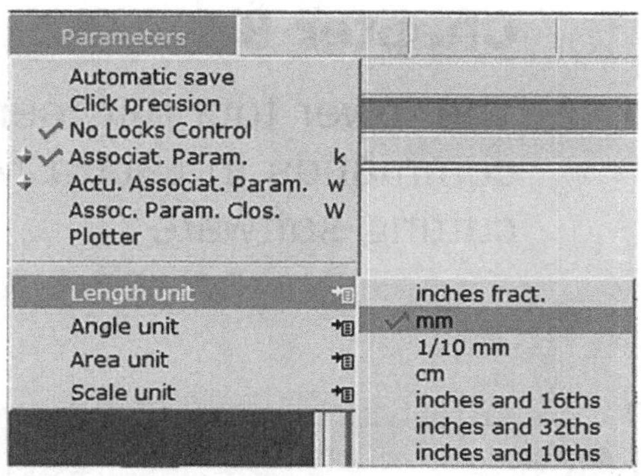

8.40 Length unit choice window

8.12 Configure menu

This menu is used for customising Modaris menus and is mostly for the advanced user (Fig. 8.41). Icon/text is used to select the normal working display as either text or graphics. See Chapter 6, Section 6.1.3 for more details.

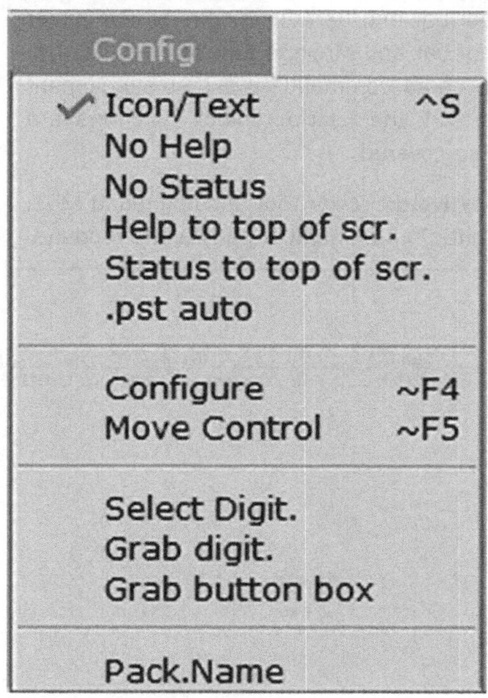

8.41 Configure menu

8.13 Tool menu

This menu is used when customising Modaris menus and is for the advanced user (Fig. 8.42).

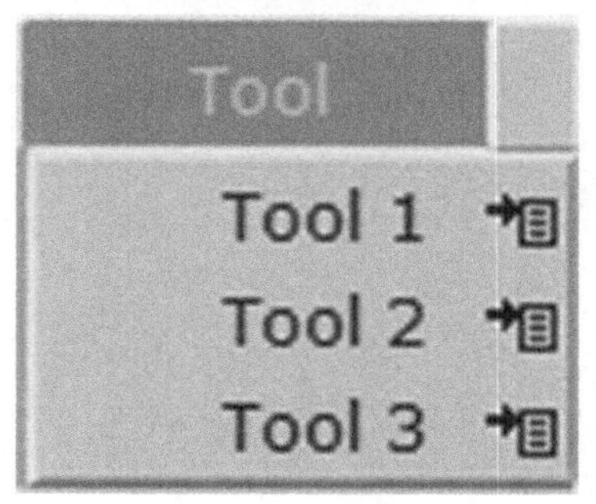

8.42 Tools menu

8.14 Help menu

This field allows access to the full online, on-screen Lectra help menu. Select and follow the prompts (Fig. 8.43).

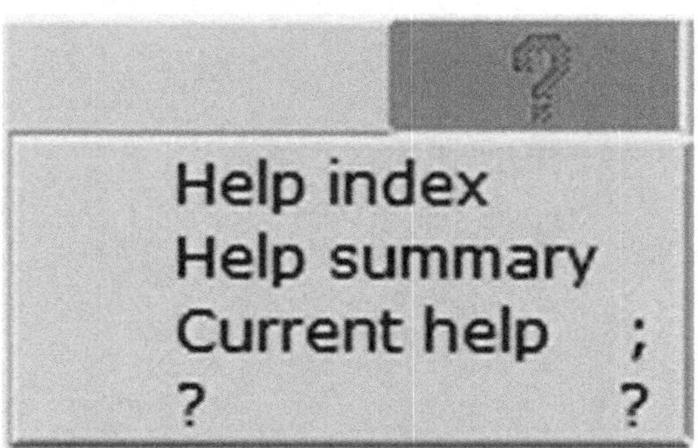

8.43 Help menu

Chapter 9

The lower tool bar menus and keyboard commands in Lectra Modaris pattern cutting software

Abstract: This chapter explains the lower tool bar and its connections to the upper tool bar and Modaris function menus. The keyboard command keys that supplement or vary the Lectra toolbox functions are also covered.

Key words: lower tool bar, command keys, Shift, Control, number pad, Lectra Modaris.

9.1 Lower tool bar

The upper half of the tool bar displays the current options in a selection of key functions (Fig. 9.1). The lower half is a series of on/off buttons for the most essential display options. All of these command fields are available in either the function menus F1–F12 covered in Chapters 6 and 7, or the upper tool bar menus covered in Chapter 8 where full explanations will be found.

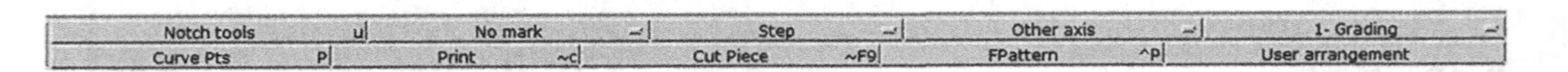

9.1 Lower tool bar

9.2 Notch tools

This field allows the selection of notch type. However, in practice, once this has been chosen, it becomes a default option and is rarely changed (Fig. 9.2).

9.2 Notch tools

9.3 Marks (F1 – Relative point)

This field displays the currently selected option (Fig. 9.3). It also allows the selection of alternative drill-hole graphics via its pop-up menu, see Chapter 6, Section 6.2.3 Relative points.

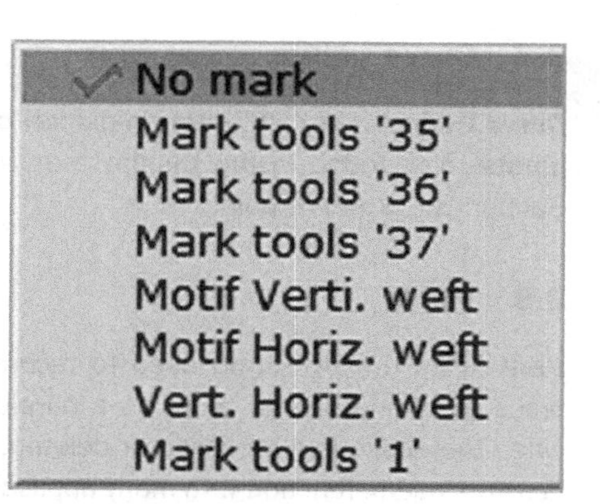

9.3 Marks tools

9.4 Corner tools (F4 – Corners)

This field displays the currently selected corner type (Fig. 9.4). It also allows the selection of alternative corner types via the pop-up menu, see Chapter 7, Section 7.2.6 Corner types.

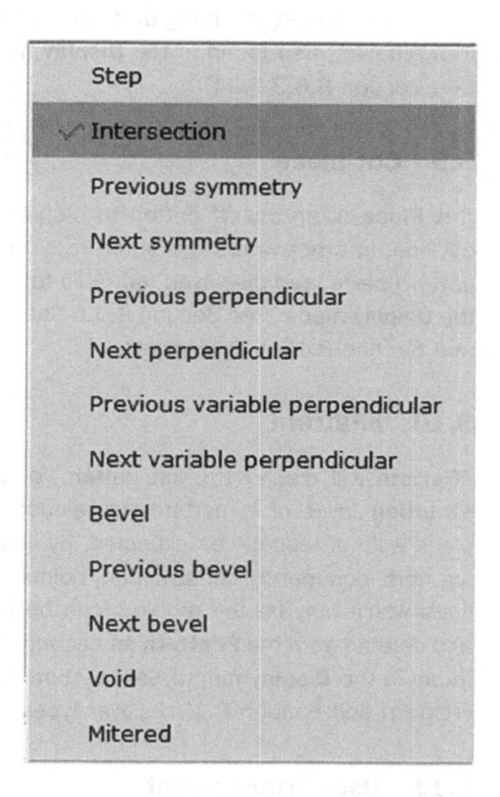

9.4 Corner tools

9.5 Axis (F4 – Axis)

This field displays the currently selected axis type (Fig. 9.5). It also allows the selection of alternative axes via the pop-up menu, see Section 7.2.4.

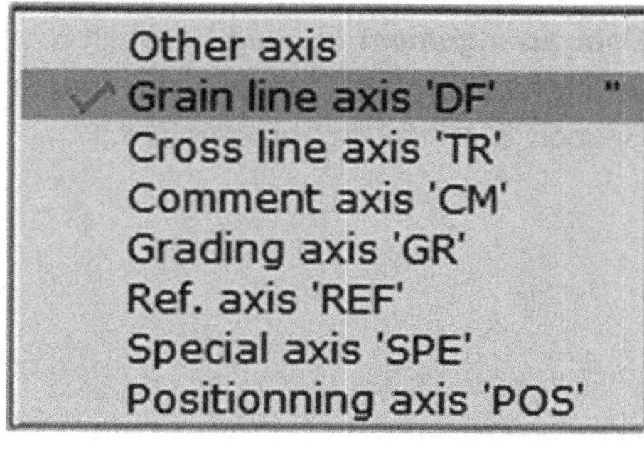

9.5 Axis

9.6 Grading

This field displays the chosen grading option (Fig. 9.6). The default setting is **1-Grading**, which is suitable for most situations. Special grading (**2-Spec. Grad. 1**) and (**3-Spec. Grad. 2**) are advanced grading functions.

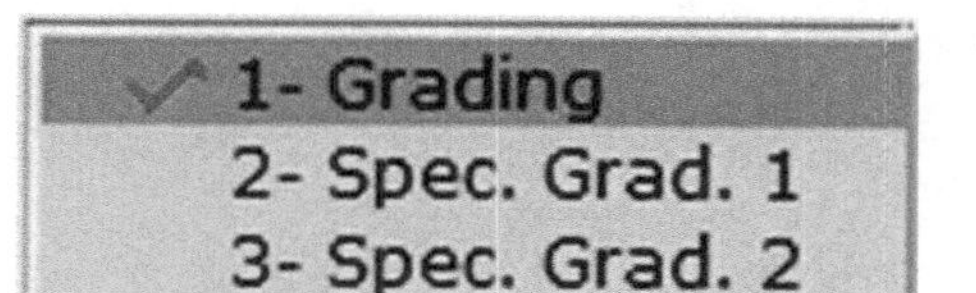

9.6 Grading

9.7 Curve points

Curve Pts is an on/off button to display **curve points**. Also found in the **Display** menu. See Section 8.6.2 Curve points.

9.8 Print

Print is an on/off button used to record the piece lines and display them as a purple line when the active line is moved or deleted. This is a very useful function with many applications and, once you get the hang of it, you will find it invaluable. Also found in the **Display** menu. See Section 8.6.3 Print.

9.9 Cut piece

Cut Piece is an on/off button to display the cut line, in other words the outer edge of the pattern piece, and displayed red. Also found in the **Display** menu. See Section 8.6.6 Cut piece and Section 8.10.1 Layer types.

9.10 FPattern

FPattern will display the flat pattern, or **construction** layer of a pattern. Occasionally a piece will not respond as expected. By activating this command, construction points and lines which may be the problem can be seen and deleted from the **FPattern** as needed. Also found in the **Display** menu. See Section 8.6.4 FPattern and Section 8.10.1 Layer types.

9.11 User arrangement

User arrangement is used to retain a specific arrangement of the pieces on screen. See Section 8.4.7 for full explanation.

9.12 Control keys RHS

These are the keys placed around the letter and number keys. Some work in exactly the same way as for any computer program, some have a specific action for Modaris.

9.12.1 Insert

This is an advanced function for customising function positions. It is not covered in this book.

9.12.2 Home

This will bring a single sheet to fill the screen. First click on the sheet, then press **Home**. **Current sheet**, located at the top of the coloured toolbox menus does the same thing.

9.12.3 Page up / down

This allows you to scroll between successive sheets. Useful when running through all of the sheets to check grain lines, text boxes etc.

9.12.4 End

This is a movement key allowing you to rearrange pieces on the desk to your requirement. Select a sheet with the mouse and press **End**, then left click and move the sheet around on the screen. Click again to place the sheet where required. Note that, when you press **J/j** or **8** again, the pieces will lose their arrangement. To maintain a specific arrangement see Section 8.4.7 Arrangement record.

9.12.5 Delete

This is an alternative to the **F3 – Delete** function. Select the item to delete with the mouse, and press the **Delete** key.

9.12.6 Enter

Besides the normal use in typing text and completing functions, this key is also used to **zoom**. Press **Enter** and a magnifying glass icon will appear. With the mouse **click and drag** to select a specific area to enlarge.

9.12.7 Full stop or period (.)

This key allows movement while in a function. It works like the horizontal and vertical scroll bars, but means you can move a sheet without quitting the active function.

9.12.8 Backspace

This activates the scale ruler. Press again to remove. Also found in the **Display** menu. See Section 8.6.13 Scale.

9.12.9 Arrow keys

These allow movement between fields in the **dialogue boxes**. See Section 6.1.1 Dialogue box.

9.12.10 Space bar

The **Space bar** is a choice bar. This bar (which has no alternative method) allows you to make a choice when more than one option is available in a function. As examples:

- when using **F1 – Align 2 points** – is it the ***x*-axis** or the ***y*-axis**? Choose with the **Space bar**;
- when using **F8 – Walking pieces** – which piece walks which, and in which direction? Choose with the **Space bar**.

Many functions have a multiple choice of action or direction. Press the **Space bar** to scroll through the options.

9.13 Control keys LHS

9.13.1 ⭾Tab

This is the shortcut to insert an existing and saved model or pieces into the open model. It is also used to access certain **dialogue boxes** when customising functions. See Section 8.2.3. Insert model.

9.13.2 Caps lock

This is used in the normal way for text and with no extra function but be aware that most shortcuts are case sensitive.

9.13.3 Esc

This is used to exit a function before its completion.

9.14 Shift

There are a number of combinations using the **Shift** key.

- **Shift** combined with a function will vary that function.
- **Shift** + **F1 – Add point** will make it a curved point.
- **Shift** + **F1 – Straight** will create the new line at a right angle to the existing line.
- **Shift** + **F3 – Delete** will also delete any **characteristic points**.
- **Shift** + **F8 – Marry** will flip the piece over.
- **Shift** + **right click mouse** allows multiple selections of lines and points. Remember to release **Shift** before completing the action.
- **Shift** + **Sheet – Delete** will delete the construction layer (grey sheet) at the same time.

9.15 Control (Ctrl)

Ctrl combined with another key or function will vary that function. There are a number of combinations.

- **Ctrl** + **F1 – Straight** will keep it on the true horizontal, vertical or 45° angle – use the **Space bar** to scroll through the options.
- **Ctrl** + **F3 – Reshape** will keep the point on the true horizontal, vertical or bias – use the **Space bar** to select which.
- **Ctrl** + **A/a** will select all pieces, which is useful when re-numbering a whole pattern. See Section 8.3.2 Rename.
- **Ctrl** + **S** will change the right-hand command column from words to icons, and back again.
- **Ctrl** + **W/w** will redo the last command
- **Ctrl** + **Z/z** will undo last command, and is not just for mistakes. By combining **Ctrl** + **Z/z** with **Print** this can be used like a tracing function. **Marry** a piece, activate **Print** and then use **Ctrl** + **Z**. The piece is **divorced** but the outline remains.

9.16 Number pad

1 will make all of your lines disappear, leaving only the coloured silhouette of the piece. This is in fact the **construction layer** of the sheet.

Strike **2** and they will reappear. Easily done when meaning to press **F1** or **F2** but miss. See Section 8.10.1 Layer types.

7 and **8** are used when you want to work with one piece or a small selection of pieces in a separate window. See Section 8.4.4 Sheet Sel.

Chapter 10

Converting files using the Lectra interoperability function

Abstract: Data exchange between the different pattern cutting software systems is vital and Lectra makes it possible to convert from one file type to the other using a function called **interoperability**. This chapter takes the user through each stage of the procedure to both import and export patterns as well as convert to the appropriate file type.

Key words: interoperability, alias, file conversion, Mdl, DXF, AAMA, PDF, import, export, zip files, Lectra Modaris.

10.1 Introduction to the file conversion procedure

Lectra is one of a number of pattern-making software systems that includes Gerber Technologies, assyst and Pad, among many others, all with different file types unique to their specific system. Data exchange between the different pattern cutting software systems is vital because patterns and layplans are now emailed between design studio and factory and around the globe between one company and another, often using different software systems. Lectra makes it possible to convert from one file type to another using a function called **interoperability**.

There are a number of file types available in interoperability but the ones I most commonly use are AAMA (American Apparel Manufacturers Association) and DXF (drawing exchange format). Ask the receiver of your pattern files what format they would like and also ask them to check that the files can be opened and used. If they are unable to open the file, try again with a different format.

Because of the continuing development of software, with a number of versions now in use, it sometimes takes a few attempts to get the whole procedure just right. Ensure you make a note of which factories or suppliers use which file exchange format for future reference.

10.1.1 File types

AAMA is a DXF type file that retains the complexity of the whole pattern. Once converted and imported by another system, the model can be used exactly as in the Lectra system. For example, it can be amended, graded and used in a layplanning program.

DXF files can be opened in illustration programs, in which case they are for the outline shape only, and no further work can be carried out on the pattern. They become an illustration rather than a pattern but can be kept to true full scale. Designers can use this for developing prints or placement embellishments.

PDF files (portable document format) are useful for attaching to an email as a small-scale illustration of a pattern or piece to send to someone who doesn't use specialist pattern-making software, but may need to see the pattern shape such as a Garment Technologist. This particular conversion takes place in JustPrint, see Chapter 5, Section 5.4.2. In the **Activity management** field (see Fig. 5.17), select **Model-PDF**. Select the required model or piece as explained in Section 5.2 and when sent to plot it will be converted to a PDF file. Save and print or email as needed.

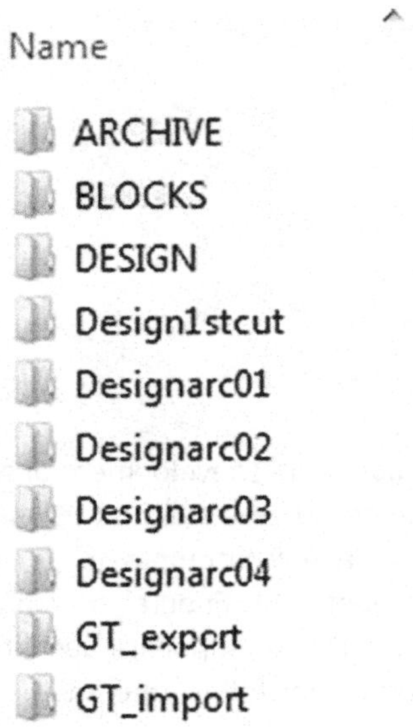

10.1 *Folder names*

Name

ARCHIVE
BLOCKS
DESIGN
Design1stcut
Designarc01
Designarc02
Designarc03
Designarc04
GT_export
GT_import

10.1.2 File conversion folder preparation

It is advisable to create separate folders in which to save the various file conversions. I find it helpful to have an import folder, in this example named **GT_import**, and an export folder, here called **GT_export**, for each file type. **GT** stands for Gerber Technologies (Fig. 10.1).

Within the import and export folders, each pattern then needs to have its own folder to contain all of the separate files associated with that pattern. Converting to other pattern-making systems often requires additional files for text information or grading rules. This will happen automatically when the file conversion takes place. To keep things simple and consistent, name the new folders created within the **GT_export** and **GT_import** folders with the same name as the original Modaris or importing file name. Because the files will have different file extensions you will not confuse them.

This chapter will explain the import and export procedure using AAMA, but the principle is the same for most types. Gerber import procedure is slightly different if the user receives a zip file.

10.2 Export procedure

This example will use the Modaris file name **A0000**, containing a pencil skirt design, and will be converted using **AAMA**.

10.2.1 Export preparation

When a pattern is ready for export conversion, in this example model number **A0000**, make sure it is saved into the normal folder location and have it visible on screen in Lectra Modaris format.

- Leave it on screen while you perform the conversion procedure.
- Conversion will take place in the **GT_export** folder.
- Name the new folders created within the **GT_export** folder with the same name as the original Modaris file name.

10.2.2 Create a new folder

Go to **File – Open**. You then need to follow the steps outlined below.

- In the **Open model file** window select the **GT_export** folder from the **look in** field using the drop-down menu.
- Create a new folder within the **GT_export** folder by clicking on the **Create new folder** icon on the right. Alternatively **right click** anywhere on the right-hand side of the screen and select **New** and **Folder** from the pop-up menu.
- Rename the new folder **A0000**.

Having created this folder, **cancel** the window and return to the Modaris screen with the Lectra format pattern.

10.2.3 *Activate export*

With your original pattern, in Lectra format, still on screen, select **File – Export** and wait for a few moments while the function opens (Fig. 10.2).

The **interoperability** window will now open (Fig. 10.3).

10.2.4 *Select format*

The **interoperability** window will have a list of different file formats with a [+] sign, folder icon and the format name. **Right click** on the actual format name **AAMA** (not the [+] plus sign or the folder icon) and two options will appear:

- New alias
- Properties

Click on **New alias** . . . and the **new alias** window will appear (Fig. 10.4).

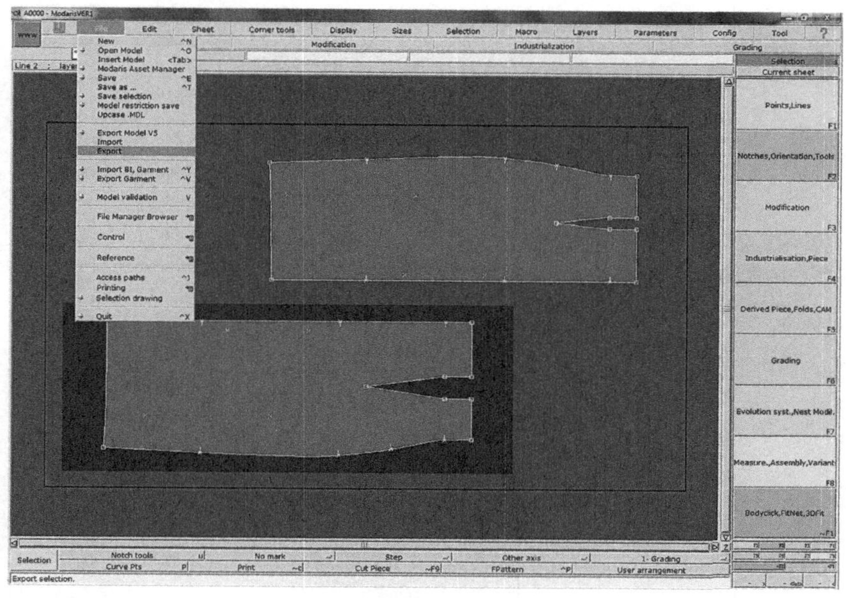

10.2 File – Export

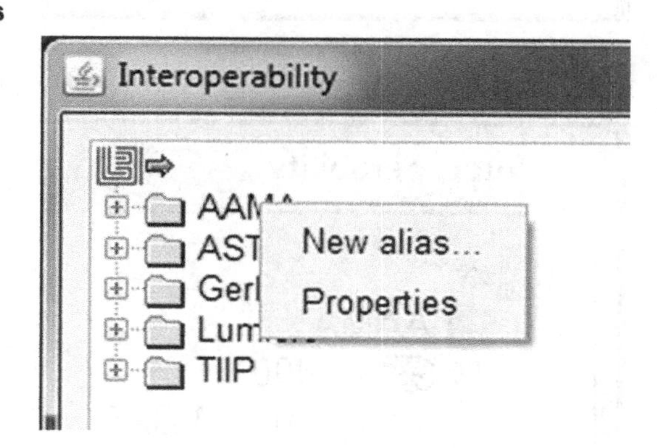

10.3 Interoperability window

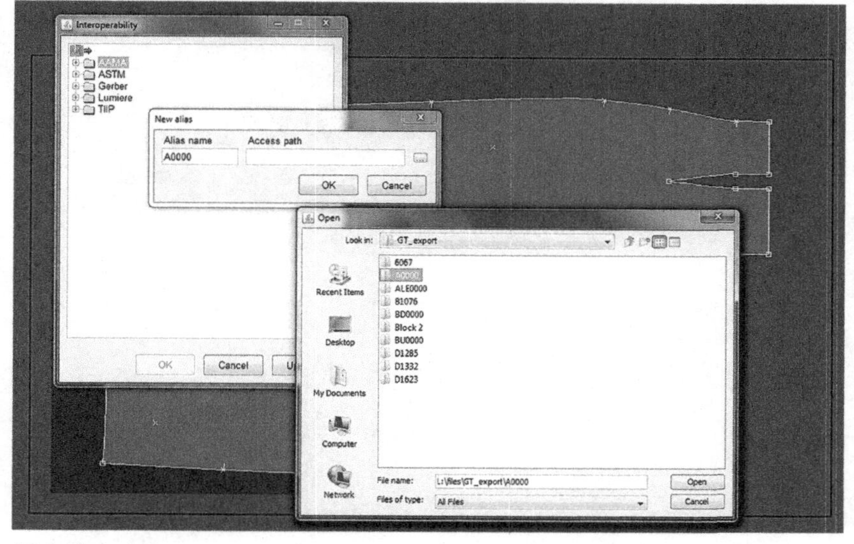

10.4 Alias window

10.5 Completed alias window

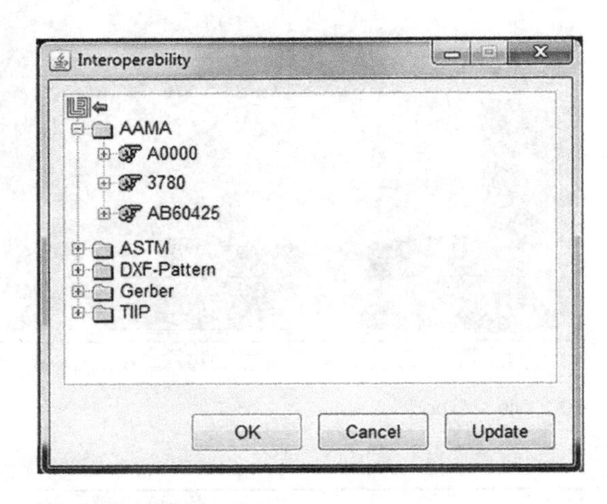

10.6 Interoperability window displaying aliases

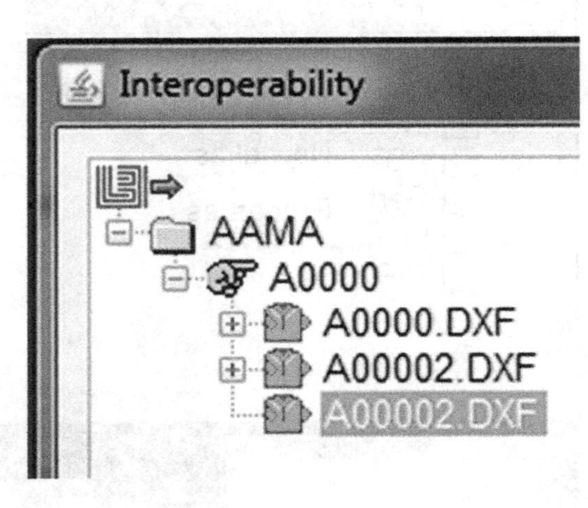

10.7 Alias folder displaying files

10.2.5 *Create a new alias*

Use the following instructions (Fig. 10.5).

- In the left-hand field, enter the **alias name**, which is the same as the folder name created in 10.2.2, in this example **A0000**.
- To complete the right-hand field, the **access path**, use the search button . . . to locate the folder you made in 10.2.2: **Lectra/files/GT_export/A0000**.
- Select **Open**, and be aware that this actually closes the window (Fig. 10.4).
- The **access path** field will now be complete, so click OK and this window will close (Fig. 10.5).

10.2.6 *Interoperability*

You will now be left with the **interoperability** window showing all of the aliases that have been created with a [+] plus sign and a finger pointing to the various pattern numbers (Fig. 10.6).

- Left click on the [+] plus sign of your **alias** and all of the pattern files relating to your pattern will be listed. They will have little garment-shaped icons beside them, and note that there may be more than one file (Fig. 10.7)
- Click **OK**

This will activate the conversion in the required format into the folder created in **GT_export**. This may take a moment or two and there will be nothing on screen to tell you that this has occurred. However, if it is not successful, a dialogue box will state that there has been an error. In this case start again making sure that all the steps are followed.

10.3 Confirming conversion has taken place

Use the following instructions (Fig. 10.8).

- Using **Windows Explorer** locate the folder created in 10.2.2
- Click to open the folder and inside will be the converted file formats ready for export.
- Be aware that it does not look the same as Lectra pattern files and may be composed of a number of files with different file extensions. This is the reason to have a folder to keep all the related files together in one place.

This export procedure will generate the DXF file (pieces) and the rul file (grading). When sharing this model, both files should be included, otherwise the grading will be missing. However, when sharing for an illustration software package, only the DXF file is required.

10.4 Email attachments

Use the following instructions.

- Create new email and attach the files in the normal way using the **Insert** button.
- Note that you cannot attach the folder but will need to attach each of the files held within the folder.
- Request that recipient confirms delivery and can open files.

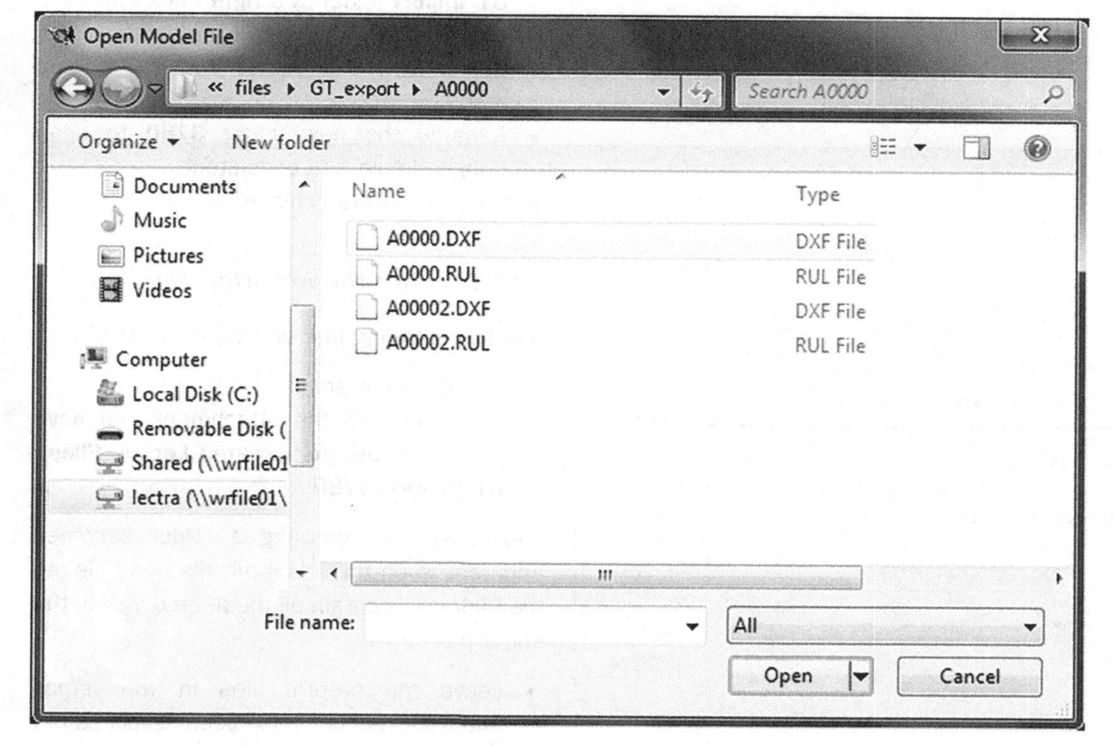

10.8 Export folder containing converted files

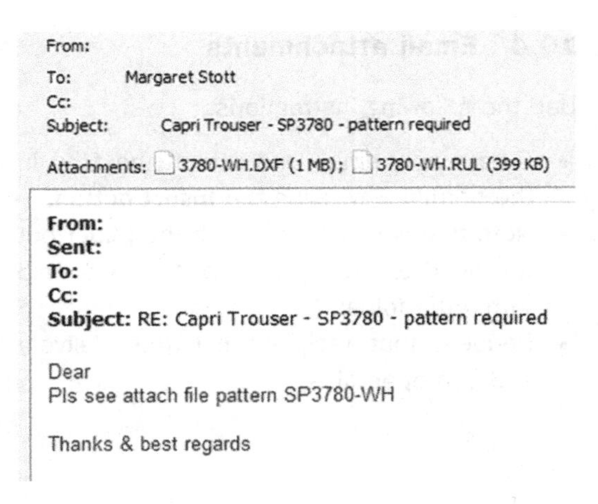

10.9 Email with attached files.

10.5 Import procedure

The key thing to note about the import procedure is that the pattern file or files received need to be saved and converted before being imported into a new Modaris model. This example will use the Modaris file name **3780**, a capri trouser design and will be be converted using **AAMA**. However, the method is the same for all file types.

10.5.1 Import preparation

When a pattern is received by email, it will appear as an attachment, or series of attachments in the email (Fig. 10.9). It may even be a **zip** file. Before doing anything with these files, create a **new folder** in which to save the files. This is efficiently done through **Windows Explorer**.

- Open **Windows Explorer.**
- Follow the pathway **Lectra / Files / GT_import** and create a new folder within the **GT_import** folder by a **right click** anywhere on the right-hand side of the screen.
- Select **New** and **Folder** from the pop-up menu.
- Rename the new folder **3780** to keep things simple and consistent.
- Close Windows Explorer.

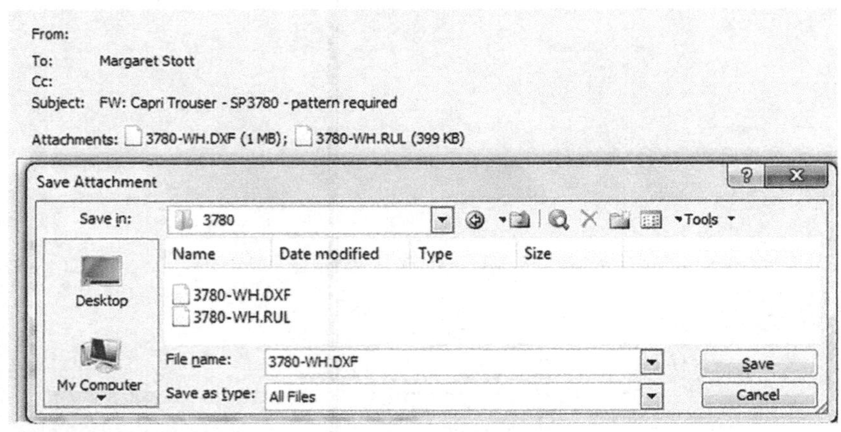

10.10 Saving attached files

10.5.2 Save the imported file

Use the following instructions (Fig. 10.10).

- Return to the email.
- Right click on the attachment and **save** into the folder just created **Lectra/Files/GT_import/3780.**

The reason for creating a folder becomes apparent when there is more than one file, as the folder will contain all the related files in the same place:

- Leave the pattern files in your email until conversion has been successfully completed.
- Close the email file.

10.5.3 Zip file

10.11 Zip file icon examples

Before the import can take place any zipped files need to be unzipped (Fig. 10.11). This applies to all formats, except for Gerber. Gerber has the option to export tmp files saved inside a zip folder. These files should not be unzipped. Instead the unzipping should be performed using the interoperability tools included in Modaris. There are a number of ways to do this, depending on the type of zip file and your software but the general principle is as outlined below.

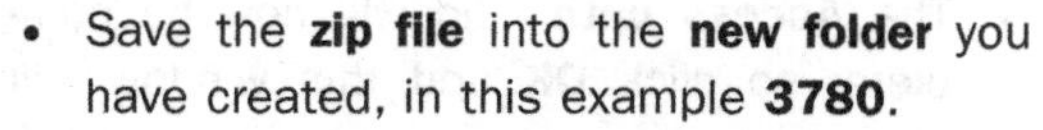

- Save the **zip file** into the **new folder** you have created, in this example **3780**.
- Double click on the **zip file** and the **zip window** will open.
- Select **Extract** and another window will open.
- Select **OK** and the unzipping will take place. This may take moment or two.
- Close the zip window, and then close **Windows Explorer** if necessary.

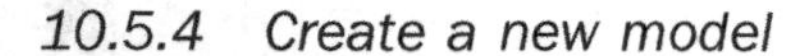

10.5.4 Create a new model

Use the following instructions (Fig. 10.12).

- Open Modaris.
- Select **File – New** and name it **3780**.
- **Save as** into the **GT_import** folder.

This folder will now contain the saved attachment files from the email and the new modaris file identifiable by the file extension or file icon.

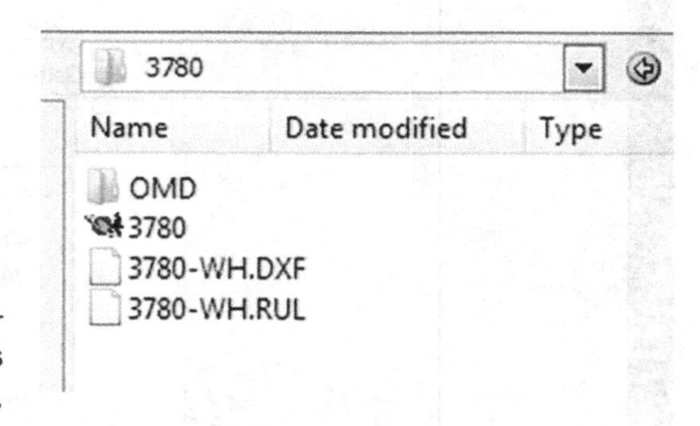

10.12 Import folder with Modaris and imported files

10.5.5 Activate import

The new Modaris file **3780** is now open on screen (Fig. 10.13). At this stage there are no sheets apart from the **model identification** sheet.

- Select **File – Import** and wait for a few moments while the function opens.
- The **interoperability** window will now open.

10.5.6 Select format and create new alias

Use the following instructions (Fig. 10.13).

- **Right click** on the actual format name **AAMA** (not the [+] plus sign or the folder icon) and two options will appear: **New alias** or **Properties**.
- Click on **New alias** . . . and the **New alias** window will appear.
- In the left-hand field, enter the **Alias name**, which is the same as the folder name, in this example **3780**.
- To complete the right-hand field, the **Access path**, use the search button . . . to locate the folder you made in Section 10.5.1 **Lectra/files/GT_import/3780**.
- Select **Open**, and be aware that this actually closes the window.
- The **Access path** field will now be complete, so click **OK** and this window will close.

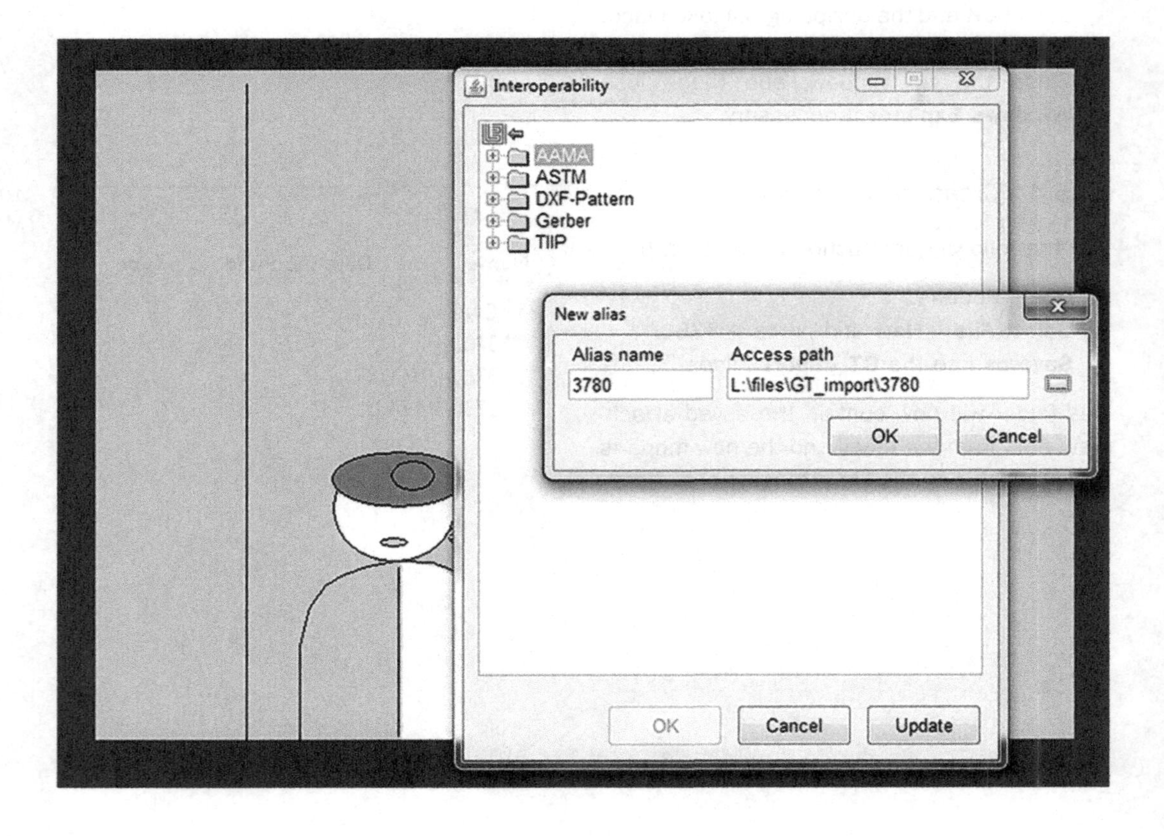

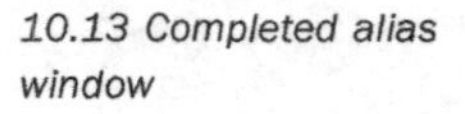

10.13 Completed alias window

10.5.7 Activate import

You will now be left with the **interoperability** window showing all of the aliases that have been created with a [+] plus sign and a pointing finger to the various pattern numbers (Fig. 10.14).

- Left click on the [+] plus sign of your **alias** and all of the pattern files relating to your pattern will be listed. They will have little garment-shaped icons beside them, and note that there may be more than one file.
- Select or highlight all of the files.
- Click **OK**.

This will activate the conversion and may take a moment or two while the imported pattern is inserted into the open Modaris file (Fig. 10.15).

Now resave the file this time containing the imported pattern file.

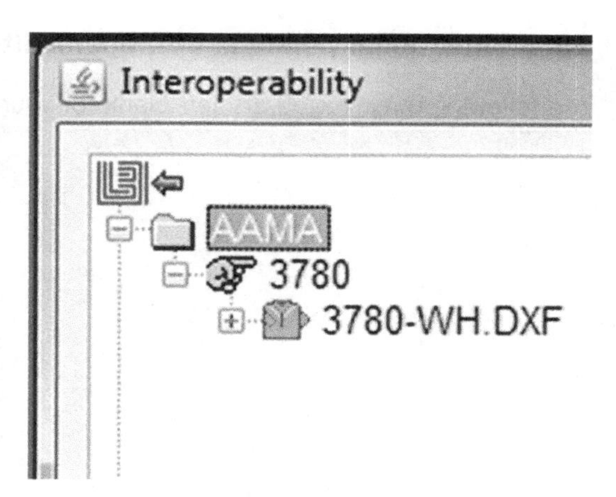

10.14 Import alias folder

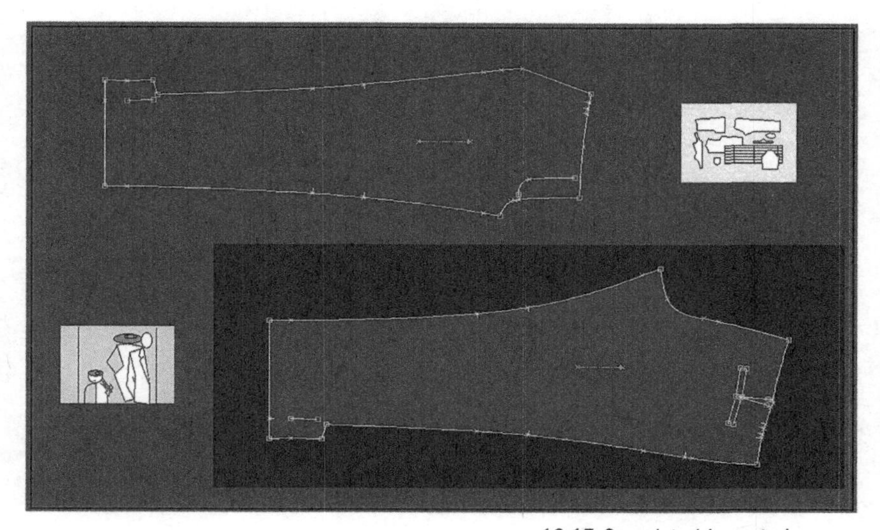

10.15 Completed imported pattern

10.6 Error messages

If either import or export has not been successful a dialogue box will state that there has been an error:

- Try again making sure that all the steps are followed.
- Try a different file format
- Call the Lectra helpline who will be happy to take you through the procedure.

10.7 Additional personal interoperability notes

The following three pages are left blank for your own notes.

Chapter 11

Creating size ranges in Lectra Modaris pattern cutting software

Abstract: The size ranges on offer in the clothing world are extensive and varied so it is essential to be able to write size range files for specific needs. This chapter explains what a size range is and why it is so important. Finally there is a guide and examples on how to create and use size range files.

Key words: size range, base size, numeric, alphanumeric, F7-Imp. EVT, Lectra Modaris.

11.1 Introduction

11.1.1 Menu F7

The Lectra menu **Evolution system**, **F7** is all about sizes, how size ranges are named, added to and changed (Fig. 11.1). The only essential function in this menu to begin with is **Imp EVT**, which attaches your chosen size range to the model.

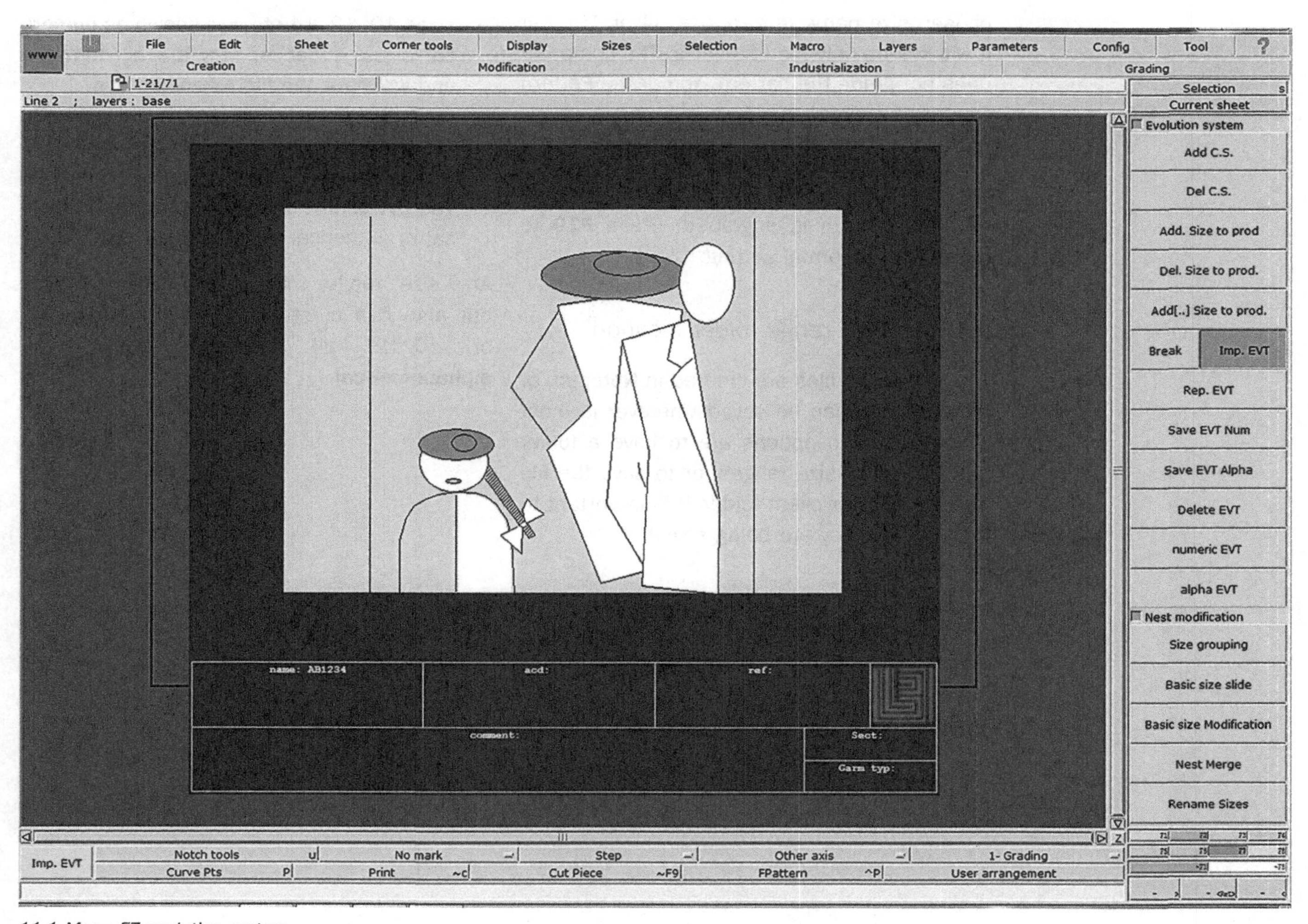

11.1 Menu F7 evolution system

11.1.2 Why add a size range?

It is important to have a size range added to your pattern right at the beginning when creating a new file, because the pattern will not plot out if it does not have a size to refer to. Other problems can arise too, and it may not be apparent that the lack of a size range is the cause. Adding the size range after completing a pattern can be a long-winded action, so it is advisable to add the size range before creating a new pattern.

11.1.3 Grading

It is also important to note that, although the size range is attached to the pattern, the pattern is not yet graded. If, without grading the pattern, a size different to the base size is selected for plotting, it will still be only the base size. The pattern will plot and have the chosen size name (e.g. a size 14 or 16), but it will still be only the base size. The computer does not grade the pattern. You or the Pattern Grader do that job.

Press **F9** and there will only be a solid orange line with no nest of grades visible, verifying that the pattern is ungraded. Press **F10** to return to the normal setting.

11.1.4 Size range folder location

The size range files are created in Notepad, or Wordpad and can be saved wherever is most convenient. Two options are to have a folder specifically for size ranges, or to save the file into a particular client folder. It is important to know where they are being saved.

11.2 Terminology

Size or **size name** refers to the individual size such as a size **12**, or **S** (small) or **10R** (10 regular).

Sizes, or **size grouping**, refers to a range of sizes. A **size grouping** for example may be sizes **8–16**, **38–44** or **SML** (small, medium and large).

The **base size** is the size that will be displayed on screen when the pattern is opened. If you sample in a **size 10**, then you need to have a **size 10** as your **base size**.

If you sample in a **M**, then you need to have **M** as your **base size**. When creating a size grouping you can define the base size to suit your pattern-making requirements.

There are two types of size ranges:

- sizes 10, 12, 14 etc. are known as **numerical** sizes – that is, they are all numbers and will have the file extension **evn**.
- sizes S, M, L etc. are known as **alphanumerical** sizes – that is, they are all letters and will have the file extension **eva** or **evt**. The **evt** extension can be either alphabetical or numerical depending on content.

Any size range that is primarily numbers but also has a letter or symbol, such 12R, or 10/12 will also be classed as **alphanumerical**.

11.3 How to create a size range

11.3.1 Preparation

Decide on the **name** of each of your sizes. This can be anything you require but there are industry standards that are best followed. Decide on the **base size** of the range. This will be displayed in white in the size range text box, and also be the size of the pattern displayed on screen when the model is opened. The graded nest, once the pattern has been graded, will display each size outline in the same colour as the corresponding number in the size range. Size 6 will be displayed in yellow, size 8 in green etc. (see Fig. 11.2).

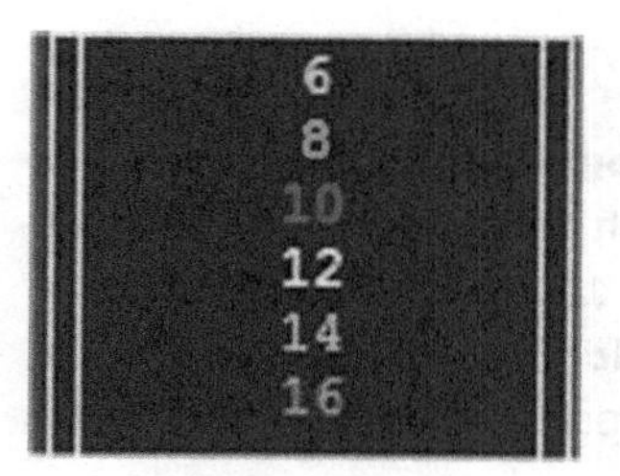

11.2 Display of a size range

Decide on the **name** of the grouping. This is the file name so it needs to be something easily identifiable. Figure 11.3 shows some examples of size range names.

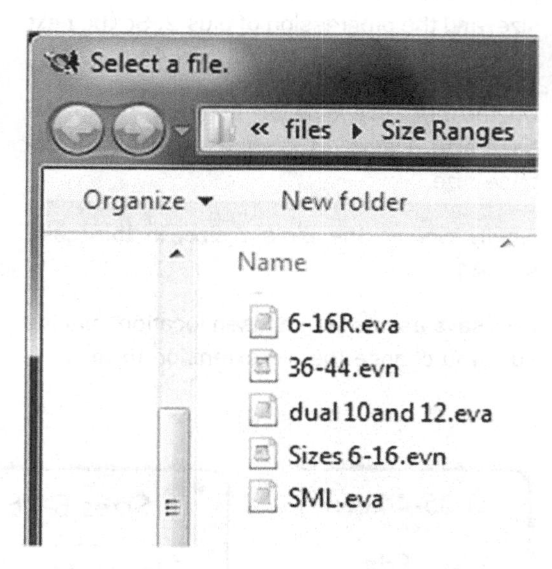

11.3 Folder list of size ranges

There are a number of ways to create size files, so if you find that one method does not suit your computer system, try an alternative method. I have found that with computer upgrades sometimes the old method no longer works.

One method is to create and name the file before it has any information contained within it. To do this, open **Windows Explorer**, and locate your **size range** folder. Right click anywhere on the right-hand side and go to **new** and then **text document**.

Now **save as** and give the file the **name** you require, completing the field with the file extension **.eva** or **.evn** as appropriate. This action will happen in Notepad or Wordpad, which will probably want to give a .txt file extension. Lectra assure me this is OK, but that there must also be the **.eva** or **.evn** in the file name as well.

Close **Windows Explorer**. Go to **Start – Programs – notepad or wordpad** and open the newly created file, then follow the instructions as shown in Section 11.3.2 and Section 11.3.3.

A second method is to save the files after entering the size range information as explained in Section 11.3.2 and Section 11.3.3.

11.3.2 Writing numeric sizes

Go to **Start – Programs – Accessories** and then **Notepad** which opens a white screen (see Fig. 11.4 and Fig. 11.5). Start at the top line and type in **numeric**. On the next line type the smallest size, a space, and the size of the step between sizes.

Figure 11.4 shows size **36** as the smallest size, and the progression of plus 2. So the next size will be **38**, then **40**, **42** and **44**.

On the next line type an asterisk * and the **base size**. Figure 11.4 shows **size 40** as the base size.

Finally type in the largest size, in this case size **44**.

Now **save-as** into your chosen location, making sure you change the file extension to **.evn**.

Note that, for numeric size ranges, not every size needs to be entered.

11.3.3 Writing alphanumeric sizes

Go to **Start – Programs – Accessories** and then **Notepad**, which opens a white screen. Start at the top line and type in **alpha**. Remember that **alphanumeric** sizes need to have every size entered onto separate lines. On the next line enter the smallest size, followed by each size name on the subsequent lines.

Indicate the **base size** with an asterisk *. Figure 11.6 shows all letters. Figure 11.7 shows numbers and letters, and Fig. 11.8 shows numbers and a symbol, in this case a forward slash.

Now **save-as** into your chosen location, making sure you change the file extension to **.eva**.

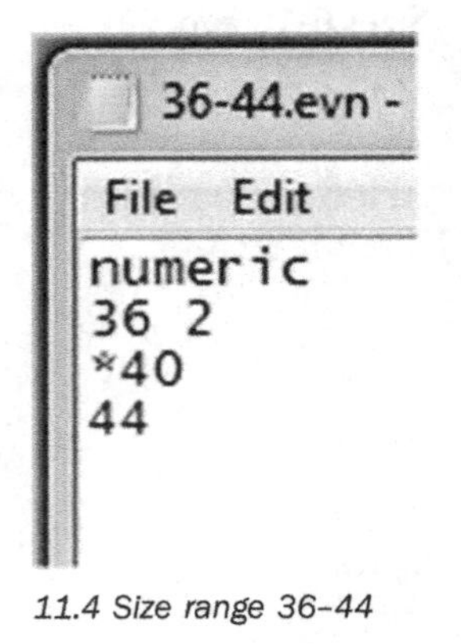

11.4 Size range 36–44

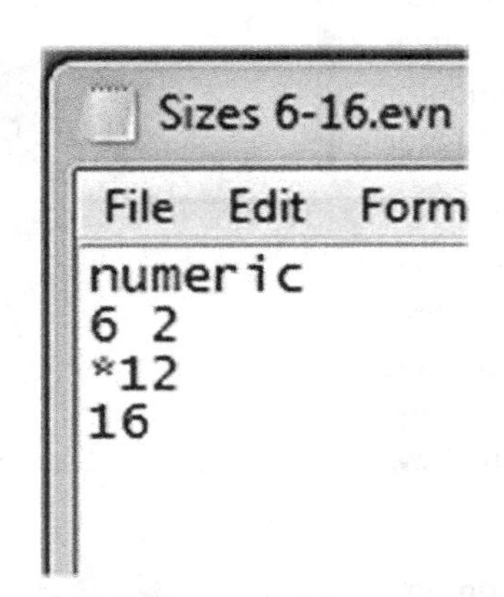

11.5 Size range 6–16

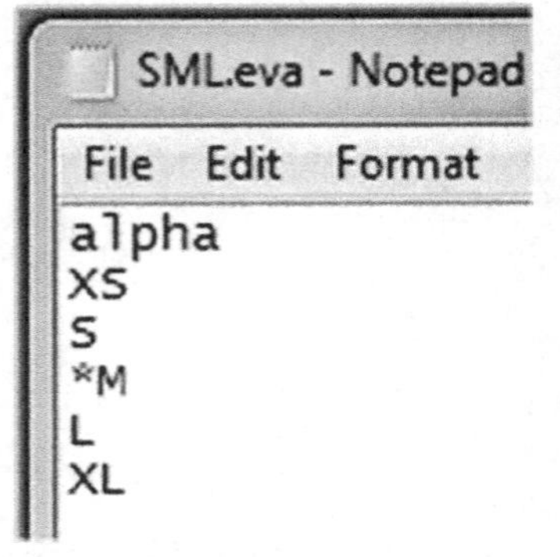

11.6 Size range XS–XL

11.7 Size range 6–16R

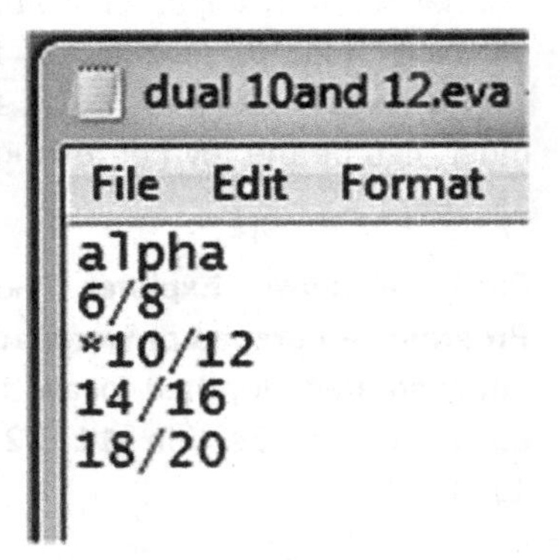

11.8 Size range dual 10 and 12

11.4 Attaching a size range to a model

When creating a **new model**, double click in the centre of the **model identification** page to access the size range files, select the required size range and press **Enter** (see Fig. 11.9).

The selected sizes will now be assigned to each subsequent sheet created in the model file (see Figs. 11.10–11.14).

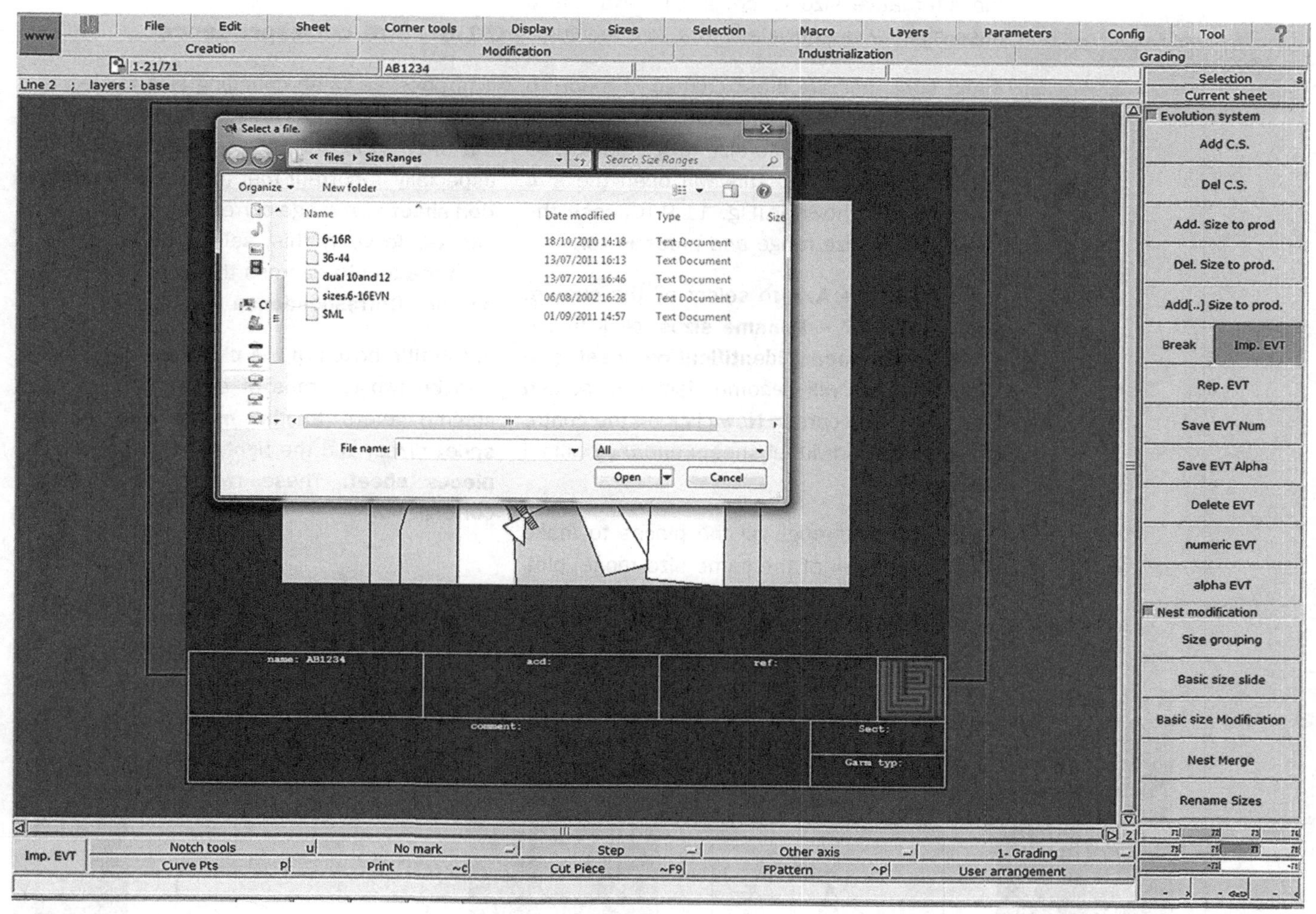

11.9 Accessing the size range folder

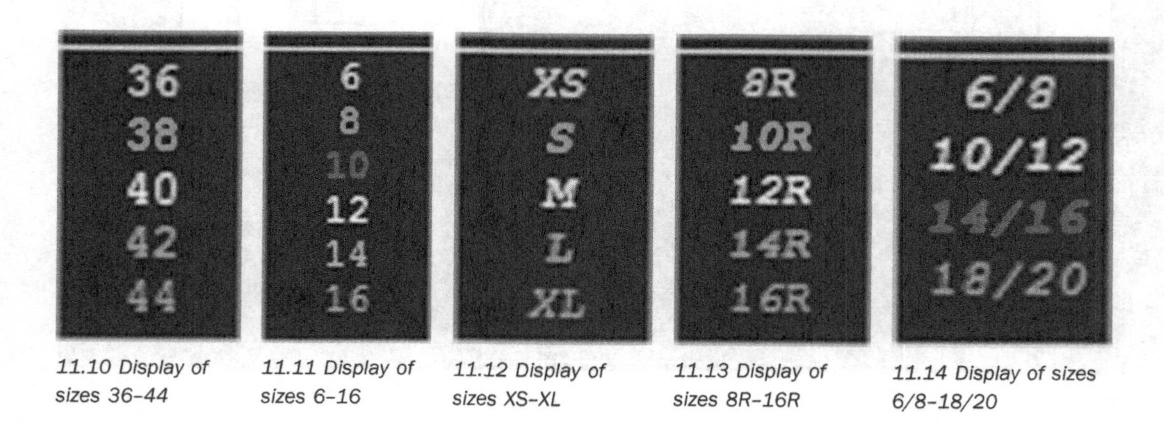

11.10 Display of sizes 36–44

11.11 Display of sizes 6–16

11.12 Display of sizes XS–XL

11.13 Display of sizes 8R–16R

11.14 Display of sizes 6/8–18/20

11.5 How to change a size range

Sometimes a size range already assigned to a pattern needs to be changed. An example is to change from a 10/12/14 to a S/M/L.

First make sure that the base size in the existing range is an equivalent to the base size in the alternative size range. In this example a base size 12 is equivalent to a size medium.

Make sure that the firest change is made to the **model identification sheet**. Double click with the cursor over the size range display, as shown in Fig. 11.11. This will open the size range file as shown in Fig. 11.9. Choose the relevant new size range and press **Enter**.

Now use **Ctrl** + **A/a** to select all the sheets. Then select **F7 – Rename sizes**, click in the centre of the **model identification sheet**, and the size range will become visible on screen just next to the cursor. Now click on the centre of any of the individual sheets and they will all change.

Finally, check through all the pieces to make sure they are all of the same size range, plotting problems may occur if they are not the same size range. For any sheets that have not changed, select **F7 – Rep.EVT** and then click on the **model identification sheet**, on top of the size range, and then on the unchanged sheet, on top of the size range. The size range will now change.

11.6 Size correspondence

It is possible when changing size ranges that some of the sheets are overlooked and that the sheets do not correspond correctly. It is especially important that the **model identification sheet** size range corresponds to the **piece sheets**. To verify this, select the **sizes** menu from the tool bar across the top of the screen and tick **Correspondence** (Fig. 11.15).

In the title boxes of the **piece sheet** there will now be two columns of sizes. The left hand column refers to the **model identification sheet** sizes, and the right-hand column to the **pieces sheet**. These two columns should correspond.

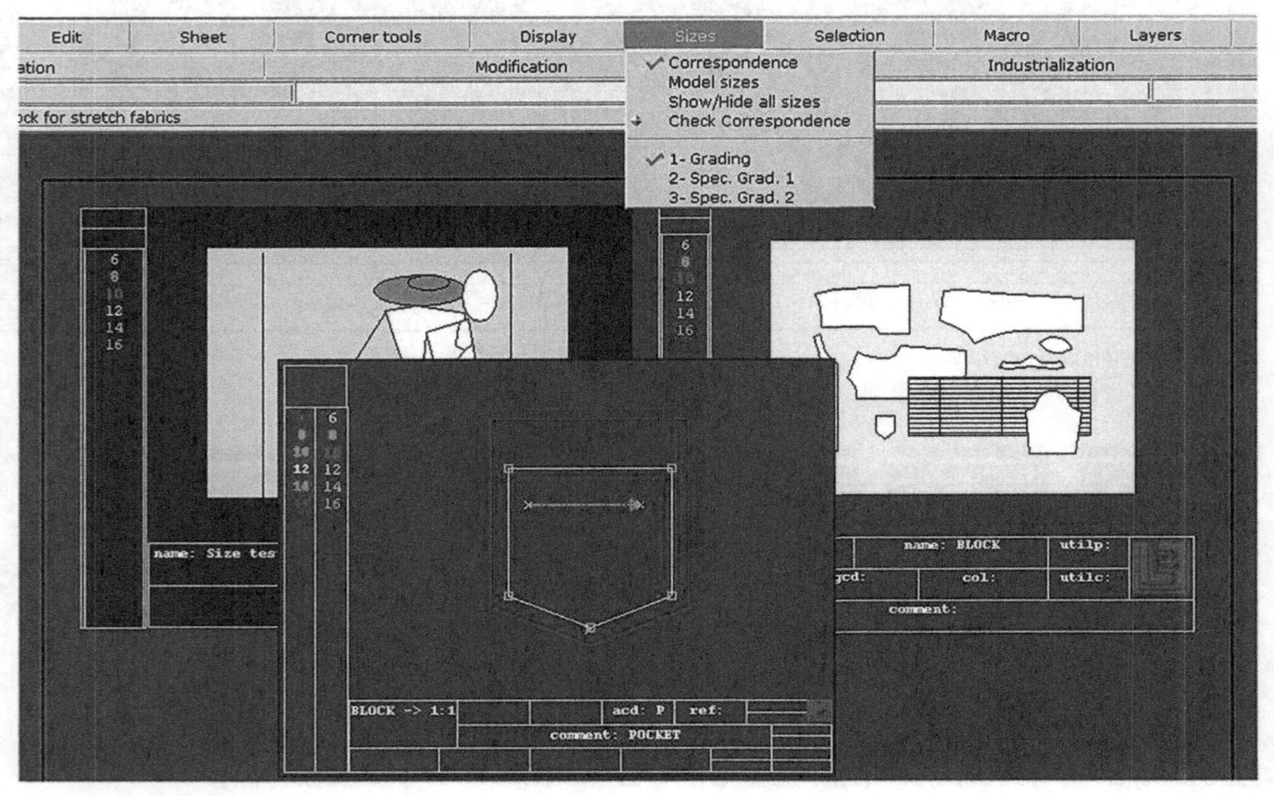

11.15 Size correspondence

Chapter 12

The importance of measurements and size charts for pattern cutting

Abstract: Pattern making is a creative art requiring intuition and judgement but it is also a science demanding rigor and methodical exactness. This chapter explores the history and development of standard size charts for industry. It looks at where the data have come from, who complied the charts and why they are so important.

Key words: pattern drafting systems, scientific surveys of body measurements, changing body shape, influences on measurements, size charts, Lectra Modaris.

12.1 Introduction: the importance of measurements for clothing

The previous chapter discussed creating size ranges in Modaris. This chapter takes a broader look at the general issues involved in garment and size measurements. It is important to be aware of the strengths and weaknesses of the measurements you will encounter.

If you are going to make patterns for clothes it is necessary to have at least some basic measurements. We can assume that measurements have always been used in one form or another. The tape measure as we know it today has evolved from necessity with Tailors and Dressmakers traditionally creating their own versions. These home-made devices were quite individual, some made from parchment paper and some from cloth marked off in regular sections, and subject to stretching or shrinking.

Any recordings, therefore, on specific measurements might vary from one tape-measuring device to another. This is fine if you are making for yourself or for an individual client. In this case you can try things on, alter and, by trial and error, make a good garment. However, in commercial clothing manufacture, where the final retail outlet for a garment is unknown, and the final product is available for purchase by anyone with the resources to buy, a trial and error approach is no longer adequate.

Pattern cutting is partly a creative art. It relies on the feel or shape of the pattern pieces combined with the many practical skills learnt on the job. These skills are very hard to pass on except by demonstration or through the mistakes and happy accidents encountered upon the way. However, pattern cutting is also a science where experimentation and evaluation have led to reliable mathematical methods to achieve certain outcomes. Measurements then become another tool at your disposal. It is worth making notes of key measurements as you discover them and in time you will use them almost without thinking.

12.1.1 Body and garment measurement charts

There are two types of measurement charts. One is for the body, and the other for the garment, referred to as the garment specification. In practice they are used together. It is a part of the Pattern Cutter's task to bring the two together in creating stylish clothing that fits the human body properly.

Garment measurements are style-specific. Hence the need for specifications. What is the lowest commercially acceptable neckline? What is the smallest neckline that you can pull over your head without using an opening? How wide is a wide-leg trouser hem and what is the narrowest trouser hem that you can get your foot through? How long (or short) is a mini skirt?

These sorts of measurements are garment-specific, subject to variation due to fashion trends, fabric types and new technology, making it difficult to say exactly what they should be. What is now accepted as a low neckline was once deemed too immodest, and the addition of stretch to a fabric makes a close-fitting neckline or a narrow trouser hem width without an opening quite possible. There are many other measurements (too many to hold in your head if you are like me), for which it will be useful to have a reference chart.

Modern body measurement charts have taken a long time to compile, and with the development of ready-made clothes have become an essential reference guide for Pattern Cutters. How is a chart compiled, from what data and based on what criteria? How these measurement charts, ranging from empirical observation to scientific study, came into being has a fascinating history.

12.2 The development of clothing patterns

Until the early 19th century, clothes were made individually for specific individuals and could be fitted to the wearer's requirements. Most tailors and dressmakers would also alter clothes as needed. Measurement charts as such were not crucial. The garment was tried on, pinned as needed and the alterations made. Those unable to afford the services of a Tailor or Dressmaker made their own clothes. Up until the 1970s home dressmaking was still commonplace in many developed countries, complementing bought clothes which were comparatively expensive.

There is no definitive date for the change from one to the other. It was, and still is, a gradual process varying with location and wealth. Today bought clothes are relatively cheap and fabric shops and haberdashers hard to find. Most people now buy all of their clothes, and the paper pattern for home dressmaking has become almost redundant. There is evidence of resurgence in home crafts that may reverse this trend, but that is another story.

Like most of our needs for daily living, the making of clothes was a craft skill that was passed on from one person to another, school teacher to pupil from a very early age, mother to daughter, master (Tailor or Dressmaker) to apprentice. There are not many written records of their methods or patterns until the mid-1800s. Even then the patterns that were produced were aimed at the professional Dressmaker or Tailor who still needed a certain amount of know-how to use and adapt the patterns.

The history of the paper pattern is well documented in Kevin L. Seligman's '*Cutting for All! The Sartorial Arts, Related Crafts, and the Commercial Paper Pattern*' (1996). Chapter 1 gives a detailed and well-illustrated history of the emergence of the paper pattern and is worth seeking out. Another notable publication is Norah Waugh's '*The Cut of Women's Clothes 1600–1930*' (1968) which collects together many historical examples of patterns from women's magazines, private collections and museum archives. Some were to be bought as usable paper patterns, and some as scale diagrams to be drawn out full scale by the user.

Measurements relating to the body in these early patterns are few, usually only a breast (bust) measurement. More importantly, the figures on the diagrams relate to the way in which the style will be drafted, showing the shape and the proportional measurements of one seam to another. Other instructions refer to the garment in terms of 'the petticoat is made from 4 widths of 22½″ silk, 37″ long at the centre front . . .' (Waugh, 1968, p. 306). The instructions then, were a basic drafting method for a style that the maker went on to customise to the size of the wearer.

The chapter on 'Twentieth-century dress production' in Norah Waugh's book (pp. 264–266) gives a valuable insight into the gradual cross-over from dressmaker-made garments to ready-to-wear, and the many grades in between whereby garments could be bought part ready-made for a final seam to 'fit' the wearer, to be sewn at home.

12.3 The emergence of modern clothing patterns

There were many patented systems and drafting formulas employed by tailors (and indeed still in use today), often requiring special instruments such as the tailor's proportional setsquare, or special curves for armholes and neck, as well as shapers for collars and lapels.

'*Ladies' Garment Cutting and Making*' by F. R. Morris (1934) gives comprehensive instructions for direct drafting of classic ladies' tailored garments, plus many introductory essays giving an insight to the prevailing attitudes of the time. In his very first sentence Morris says:

> We must first realise that pattern production in its entirety cannot be based upon a mathematical formula. Calculations and measurements derived from the breast size will provide a working basis for obtaining the general outline of the pattern in accordance with the figure shape, but only practical knowledge and experience of modelling a garment to the figure give a scientific method of approach for the attainment of perfection of fitting qualities. (Morris, 1934, p. 1).

Despite his warning, Morris does in fact use a set procedure for drafting a garment that requires the use of measurements taken directly from the client and inserted into the formula. Built into the draft were extra 'inlays', that is extra fabric to allow for fitting adjustments, as it was during the personal fitting session that the garment was perfected. Throughout the book Morris gives us access into the working methods of the day, the problems of dogmatic attitudes and the need to modernise outdated methods.

Interestingly Morris includes a chapter on ready-to-wear, called 'Pattern construction for the wholesale trade' (Chapter XXIV) where he states: 'Undoubtedly the future of the garment trade as a career for men and women lies in the wholesale side of the business' (Morris, 1934, p. 439). Retail at this time meant the Tailor or Dressmaker worked directly with the client, as distinct from today where retail means any shop open to public patronage. Morris clearly recognised that there was a change already in progress.

12.3.1 *Commercial paper patterns*

Ebenezer Butterick, founder of the American paper pattern company Butterick in 1861, is recognised as the first to produce standardised paper patterns in different sizes for dressmakers. Taken from their website, here is a romantic version of the possible event:

> The year was 1863. Snowflakes drifted silently past the windowpane covering the hamlet of Sterling, Massachusetts in a blanket of white. Ellen Butterick brought out her sewing basket and spread out the contents on the big, round dining room table. From a piece of sky blue gingham, she was fashioning a dress for her baby son Howard. Carefully, she laid out her fabric, and using wax chalk, began drawing her design. Later that evening, Ellen remarked to her husband, a tailor, how much easier it would be if she had a pattern to go by that was the same size as her son. There were patterns that people could use as a guide, but they came

> in one size. The sewer had to grade (enlarge or reduce) the pattern to the size that was needed. Ebenezer considered her idea: graded patterns. The idea of patterns coming in sizes was revolutionary.

It was 1866 before Ebenezer began to offer women's patterns in graded sizes, along with various publications to both market and supplement his patterns. In 1930 The Butterick Publishing Company produced '*Making Smart Clothes*', just one of a number of small books written to accompany their home dressmaking patterns. There is a chapter, 'Modern methods of fitting' (Butterick, pp. 39–49) devoted to fitting techniques, recognising that some 70 years after the introduction of size variations, the pattern is still a starting point and that, for the perfect garment, further fitting will be required during the making up.

12.3.2 A textbook for Student Pattern Cutters

'*Dress Creation*' by Philip H. Richards (1937) was specifically written for Student Pattern Cutters and Designers in training to make patterns for use by the manufacturers supplying the growing wholesale market. The change from retail, namely from making directly for a single client, to wholesale, that is to say offering ready-to-wear garments to the general public, necessitated the introduction of different sizes. Knowledge of sizing that had been intuitive to Tailors and Dressmakers, and used as and when required, now needed to become standardised and written down. Richards raises many questions about how to cut well-fitting patterns for clothes to be bought by the general public. He says very succinctly:

> At the outset it is necessary to establish certain very definite precepts. About Cutting there is no sort of magic. From time to time we hear of new and wonderful "systems" whereby "perfect fitting" garments may be produced without any "trying on", merely by taking and applying certain measurements with or without the help of charts, patent measuring instruments, etc.
>
> Against imaginary moonshine of this sort I invite you to consider why it is that women of large spending power, who obtain their clothes from luxury shops and model houses where the world's most expensive Cutters and Fitters are employed, find several fittings necessary, and why large stores and shops spend thousands of pounds every year in alterations to ready-made clothing.
>
> Miraculous claims of the nature described may always be dismissed as valueless. If the fitting of the human form were so simple as these folk say, the trade would soon lose its appeal as a remunerative vocation. (Richards, 1937, p. 9).

Later in Chapter XII Sizing, Richards continues:

> Throughout the trade there is very little agreement regarding sizing. What is generally known as a 'W' or 'women's,' size may measure anything from a 36″ to 40″ bust according to the manufacturer who has made it. Efforts have been made from time to time to formulate a Standard Size Chart, but I am pessimistic enough to think that this ideal is very unlikely to be accomplished. (Richards, 1937, p. 111).

Don't forget, this was in 1937 and we have moved on since then, but are in debt to those who recognised that guesswork was no longer acceptable and that manufacturers making ready-to-wear clothes needed a sound measurement base.

12.4 The development of sizing surveys and sizing charts

Today, there are numerous newspaper and magazine articles on the changing shape of our bodies, the fit of commercial garments and the advent of mass customisation made available through three-dimensional (3D) scanning equipment and computer aided manufacture (CAM). Each article has its own observations and opinions. Many of these commentaries make 'good copy' as they say, but many lack a real understanding of the complexity of the subject. There are, however, two important studies, which are well researched and can be reliably used to construct size charts for specific company profiles and sectors. These are the studies undertaken by:

- Philip Kunick in the 1950s and 1960s
- The SizeUK survey in 2004

Philip Kunick's '*Sizing, Pattern Construction and Grading for Women's and Children's Garments*' (1967) is a seminal work which addresses the problems of measuring and sizing in great depth. It is well worth acquiring for the background work on how size tables were researched and constructed, because the basis of his work is still in use today.

In 1951 Kunick led a team from The Clothing Industry Development Council in a national survey in the UK to measure 5000 women in a scientific manner, something that had never been done before. This in turn was edited and published in 1957 by HMSO as '*Women's Measurements and Sizes*' and in turn became the basis for the British Standards Institution (BSI) standard BS 3666 'Specification for size designation of women's wear', first issued in 1963. The most recent standard is BS EN 13402-3:2004 Size designation of clothes. Measurements and intervals.

In 2002, in response to a widely held view that the measurements of 1951 were in need of review due to the perceived change in body shape throughout the population, SizeUK utilised whole-body 3D scanning equipment to measure 11,000 people across the UK. The results of the survey were made available by subscription in 2004. Rather than publishing set size charts, SizeUK recognised the diversity of base measurements due to stature, age and ethnicity, and allows a subscriber to select data in line with its customer profile and then compile an appropriate chart using reliable statistics.

A size chart then, is a guide, a set of statistics taken from a particular set of people at a certain time and, in these two cases, well documented. Many other size charts have been developed and refined over time to determine reliable measurements to fit the average person. There are extremes, and there will always be people who just do not conform to the average, but the average is what we need to deal with here. In the end, there is no substitute for your own direct observation and you should not be squeamish about studying and measuring the human form yourself.

12.5 The influence of body shape and size on measurements

There are many influences on body shape and therefore body measurements that need to be taken into account when pattern making. These range from skeletal proportions and muscle or fat covering through to cultural inclinations and individual traits. Underwear or under garments affect posture, which in turn alters shape. Life style and occupation, age and physical ability all have an influence. Some of the most important influences are discussed here.

12.5.1 Underwear

The development of underwear through all its variety from simple bandaging to extraordinary corsets and pre-shaped bra cups can be studied in many titles specialising in lingerie. It is obvious, of course, that underwear, or the lack of it, forms the foundation for outerwear, but this key fact is surprisingly often overlooked. Corsets that drew in the waist, often quite abnormally, do not rely on any size chart, only the prevailing fashion for wasp waists. But once restricting underwear quite literally lost its hold, the real body measurement became important.

In understanding this change, let me refer to Morris (1934) again, talking about how body measurement changed during the period from about 1914 to 1937, a period when women really did stop wearing wasp-waisted corsets in favour of less restricting garments. This is the period that begins with the First World War (1914–1918), which had a huge influence on the way in which women dressed. Engaged by necessity in many jobs that had traditionally been carried out by men, they turned to more practical clothes and the wearing of trousers became more commonplace.

After the war there was no going back from the freedom of their modern clothes. About women's trouser cutting Morris (1934) says:

> Like every other section of garment cutting, trouser cutting is still carried out on the same lines as those applicable perhaps ten to twenty years ago. With men's trouser

> cutting, this may not be such a mistake, but in ladies' trousers and shorts, or for that matter any garments of this character, we have to take into account the remarkable change in the contours of the feminine figure in the last few years . . . (he continues) . . . The author has a cutting book and in it the proportionate hip size for a 26-inch (66 cm) waist girth is given as 42 inches (107 cm) The modern woman would probably measure 37 inches (94 cm) over the hips. . . . (Morris, 1934, p. 276).

Was it really that the body shape actually changed, or that underwear no longer restricted the waistline? The soft muscle and fat that covers the skeleton is yielding and can be manoeuvred into all sorts of shapes. In wearing the corset the waist was drawn in very tightly. Perhaps it was the waist relaxing that he observed rather than the hips reducing. We have seen how the waist can be restricted and similarly the breast can be shaped and positioned by the garments surrounding it. Contour cutting is a discipline in itself, with its own set of techniques and skills, expert in defining alternative body shapes.

12.5.2 Posture

How a person stands or slouches affects their body measurements. Try slouching in a corset laced up so tight it's hard to breath. I think you will find it impossible. The chaise longue, a cross between a chair and a settee, was invented to enable corseted ladies to recline gracefully in their stays. Posture is influenced by many factors, such as underwear or a person's occupation, but is also subject to fashion.

During and just after the World War II era many people had been in some sort of employment relating to the military, either as a soldier or in a supporting role, wearing uniforms and trained to have an erect posture: head up, shoulders back, chest out. Later, out of the uniform, the posture remained. Today, as we work leaning forward to our computers, our posture has become more relaxed with rounded shoulders.

From a size chart point of view, this change in posture is reflected in the change of the cross back measurement. For a bust of 36 inches (92 cm) Kunick (1967, p. 35) has the cross back measurement at 13½ inches (34.5 cm), whereas today I would use 14½ inches (37 cm) for the same nominal size. This may not seem much but it alters the cross front, the armhole position and the pitch of the sleeve. So what was seen as correct at one point in time is no longer the case at another time and will probably change again.

12.5.3 Age

The other major factor is age. The young body is still growing and developing with the proportions of one part to the other constantly changing. The bust point is high and the hips still undefined, which is precisely why fashion models are most likely to come from this group.

There will always be the exception, but generally, as a woman ages, the bust point becomes lower and the torso thickens. In later life the bone structure begins to shrink as well, resulting in loss of height. So a woman who was a size 12 in her youth, and is still a size 12 later in life may well have the same bust and hip measurements, but not necessarily in the same place, while many other related measurements may be completely different.

12.5.4 Psychology

Most women do not actually want to be dressed exactly as their body shape, but would like their clothes to make the best of their features. Some want to express individuality or be a part of a social grouping, and others want to fade into the crowd. Vanity sizing, so called because it has re-named sizes to accommodate a larger figure, is a part of the psychology of clothing. What was once called a size 12 has been renamed a size 10, which makes women very happy. To enable a woman to fit into a size 10 dress, when in fact she is a size 12, is a strong marketing device.

The area of psychology opens up a whole minefield of other considerations that we will not deal with here. Suffice to say, there are many aspects to cutting, quite apart from measurements. Let me quote Richards again:

> 'Finally I would like to warn you against too much theory. The human body is far too uncertain a quantity to reduce to a mathematical formula. It is very wise to remember that we are trying to make patterns, not to work out sums.' (Richards, 1937, p. 12).

12.6 Using and comparing sizing and grading charts

The concept of an average set of measurements that will fulfil most people's needs is challenging and many classic garments will typically have some method of overcoming this. A belt, for example, will adjust a waist measurement very effectively. Pleats or gathers in the back yoke of a shirt, or the waist of trousers and skirts will give a little more ease to the girth measurements. Cuffs might have a choice of buttons to adjust the fastening. Corsets, waistcoats and some trousers traditionally employed lace-up sections in order to be able to adjust the garment to the body. Modern jersey fabrics and the inclusion of elastane fibres in both woven and knitted materials now allow much more flexibility for garments when worn. Bespoke garments, of course, overcome all these problems by making specifically to individual requirements.

12.6.1 Size denomination

Historically, those who wrote down the methods for pattern drafting, and therefore wrote the earliest UK and American charts or instructions, chose whole imperial numbers, an understandable and practical decision. Richards (1937, p. 112) uses W as the denomination, which stands for 'Womens' as his 36-inch bust, and SW 'Small womens' for a 34-inch bust. Morris (1934, p. 17) didn't use size names but just said '(to fit a) 36-inch bust'. Kunick (1967, p. 33) has a size 12 bust at 34 inches, and a size 14 at 36 inches. Butterick (1930, p. 15) offer a 34″, 35″ or 36″ without any qualification of denomination. However, a measurement chart of necessity needs to pin down specific measurements, and there are a number of factors to be aware of when creating or reading a size table.

12.6.2 Transitional sizing

A size is not a single specific set of figures, but a range of measurements between one size and another. Being aware of the transition between one size and another will help you understand why one size chart may differ from another but, in fact, be saying the same thing. At some point a person trying on a garment whose measurements lay between two sizes will have to choose between a bit tight and a bit loose. Depending on their preference for a fit that is either 'a bit tight' or 'a bit loose', they could in fact be designated a size 10 or a size 12. Equally a person whose bust–waist–hips ratio is not 'stock size' will have to choose between 'a bit tight here' or 'a bit loose there'.

12.6.3 Imperial to metric

The change from imperial inches to the metric system has been awkward because the exact conversion does not give easy to use, rounded figures. Take a bust of 36 inches: the exact conversion is 91.4 cm while a 34-inch bust is 86.35 cm. Once converted these measurements have been adjusted to a more practical value.

12.6.4 Grading variations

The incremental step from one size to another is called the grade. A regular English grade is 5 cm or 2 inches between sizes. Not all grading systems use this amount but it is a traditional English method with a long history. Smaller sizes might use a smaller grade, 1″ or 2.5 cm. Americans use different grades for different size ranges and could be 1″ or 2.5 cm, 1½″ or 3.8 cm and 2″ or 5 cm. European grading has generally used 4 cm, something being adopted more frequently in the UK.

The following is not an exhaustive discussion on grading and its variations, which would need a book on its own, but gives a flavour of variations in grading. I am going to use a size 12 = 36 inch bust to demonstrate the possible

	Bust size	Size 10	Size 12	Size 14
A	5 cm grade	34″ = 86.5 cm	36″ = 91.5 cm	38″ = 96.5 cm
B	4 cm grade	86.5	90.5	94.5
C	Rounding up and down			
		86/87	91/92	96/97
D	Range of measurements using line A as the maximum			
		81.5–86.5	86.5–91.5	91.5–96.5
E	Range of measurements using line A as the mid-point			
		84–89	89–94	94–99
F	Range of possibilities			
		81.5–89	86.5–94	91.5–99
G	1950 sizing	Size 12	Size 14	Size 16

12.1 Size comparisons

variations that you might encounter when comparing one chart to another. Figure 12.1 line A shows a simple table for the bust measurement of sizes 10, 12 and 14. For this example the difference between any of the two adjacent sizes is 5 cm or 2 inches.

12.6.5 European grading

If we use the European practice of a 4 cm grade, in which case it is also more probable that a size 10 is used as the base sample size, we obtain line B. Using the size 10 (86.5 cms) as the starting point, we see a 2 cm difference by size 14. For the smaller sizes this isn't too much of a problem but, as the sizes become larger, there are more significant differences and the discrepancy between a 5 cm grade and a 4 cm grade becomes more pronounced.

12.6.6 Rounding out figures

Let's return to Fig. 12.1, line A and from it develop line C. Look at the inches figure and note that the metric equivalent isn't a whole number at 91.5 cm. If we are only going to work in centimetres it really doesn't matter for there to be an exact equivalent. To round up or down makes sense. So now we have 91 cm or 92 cm for a size 12.

The next point to consider is whether the 'size' measurement means to fit up to that figure, or if it means the mid-way figure. Line D shows the range of measurements using the figures in line A as the maximum, and line E the range if the figure is the mid-point. Thinking around the range of possibilities, just for a size 12 we see that it could mean the size is suitable for anyone with a bust measurement between 86.5 for a loose fit and 94 for a tight fit.

12.6.7 Comparing size designations

Figure 12.2 compares the sizing chart information for sizes 8 to 16, taken from various company websites, all accessed on the same date (24 October 2010). The variation for a size 12 is 8 cm, from 88 cm as the smallest (H&M) to 96 cm (Jaeger) as the largest. This is the same as two sizes if you are using a 4 cm grade, and of one and a half sizes for a 5 cm grade. One could argue that the two different companies have different customer profiles – H&M for young bodies and Jaeger for the more mature – but both nominally are a size 12. An alternative way to compare, and which amounts to the same thing is to look at the various size designations for 96 cm. Jaeger is a size 12, M&S and BHS are a size 14, and H&M is a size 16. Figure 12.1 line A, size 14, incidentally agrees with M&S and BHS.

Over the last ten years, since 2002 and the SizeUK survey, many companies have adjusted their sizing denomination to accommodate an increase in the 'average' size. Put simply, what

	Bust size comparisons (in cm)				
	80–87 Size 8	84–90 Size 10	88–96 Size 12	92–102 Size 14	96–108 Size 16
H&M	80	84	88	92	96
Next	84	86	90	94	99
Coast	82	85	90	95	100
M & S	82	86	91	96	101
BHS	84	88	92	96	100
Top Shop	82	87	92	97	102
River Island	84	88	92	97	103
New Look	84	88	93	98	104
Hobbs	84	89	94	99	104
Zara	86	90	94	98	102
Boden	82–86	87–90	91–94	95–99	100–104
Lands End	86–87	88–90	91–92	93–96	101–104
Jaeger	86	91	96	102	108
Whistles	82	86	91	96	101
Warehouse	82	88	93	99	104
Reiss	86	88.5	93.5	98.5	
Matches	80	84	88	92	
Browns	82–86	86–90	90–94	94–98	98–102

12.2 Size variations

was once labelled as a size 14 has been renamed to be called a size 12, sometimes referred to, as noted earlier, as vanity sizing. This has been embraced by some manufacturers and not by others. As a result our chart could be renamed (see Fig. 12.1 line G), going back to Kunick's designation where our size 12 was his size 14. If you are using a size chart that has not been influenced by SizeUK data then you may well encounter this sort of discrepancy with other more recent size charts.

In this example only the main torso girth measurement has been examined, to explain the point. In practice, many of the measurements have altered to accommodate all of the points raised with regard to fashion, underwear and psychology.

12.7 Current approaches to sizing

Look around you. It seems that people do wear the ready-made clothes and that they appear to fit quite well. I have read many newspaper articles bemoaning the fact that a size 12 in this shop is different to a size 12 in that shop, and asking why can't there be a 'standard' so that all clothes conform to the same set of body measurements. To me that's a bit like the demise of the British sausage in the face of EU regulations, allowing no local variation.

I have come to realise that local conditions are very important and need to be taken into account. Despite all of the arguments put forward regarding the fit of clothing, pattern technology has become very sophisticated, offering a range of size types that cover the majority of the population. The fact is that we are making clothes very well. Besides the Regular sizes there are Petite and Tall sizes, Plus sizes, to mention but a few, as well as different length offers available for the same size garment.

Fibre and fabric technology, combined with the development of skilled pattern and garment making, are at work making and selling clothes in a huge industry. In the end it doesn't matter which charts you are using so long as you are aware of the differences and are not afraid to adjust your methods to suit your particular customer base.

	2.5 cm grade			5 cm grade					6 cm grade				
Size	4	6	8	10	12	14	16	18	20	22	24	26	28	30
Bust	77	79.5	82	87	92	97	102	107	113	119	125	131	137	143
Above bust	72	74.5	77	82	87	92	97	102	108	114	120	126	132	138
Rib cage	65	67.5	70	75	80	85	90	95	101	107	123	129	135	141
SNP to BP	24.2	24.5	24.8	25.4	26	26.9	27.8	28.7	29.6	30.5	31.4	32.3	33.2	34.1
Waist	56	58.5	61	66	71	76	81	86	92	98	104	110	116	122
Top hip	74	76.5	79	84	89	94	99	104	110	116	122	128	134	140
Hip	82	84.5	87	92	97	102	107	112	118	124	130	136	142	148
Neck girth	33	33.3	33.6	34.2	34.8	35.4	36	37.3	37.6	37.9	38.2	38.5	38.8	39.1
Neck base	36.4	37	37.6	38.8	40	41.2	42.4	43	43.6	44.2	44.8	45.4	46	46.6
Cross back	33.4	34	34.6	35.8	37	38.2	39.4	40.6	41.8	43	44.2	45.4	46.6	47.8
Cross front	29.4	30	30.6	31.8	33	34.2	35.4	36.6	37.8	39	40.2	41.4	42.6	43.8
Shoulder	10.8	10.9	11.1	11.4	11.7	12	12.3	12.6	12.6	12.9	12.9	13.2	13.2	13.5
Nape to waist	39.2	39.5	39.8	40.4	41	41.6	42.2	42.5	42.8	43.1	43.4	43.7	44	44.3
Overarm	56.9	57.5	57.8	58.4	59	59.6	60.2	60.6	60.9	61.2	61.5	61.5	61.5	61.5
Underarm	42.5	42.5	43	43	43	43	43	43.5	43.5	44	44	44	44	44
Bicep	25.8	26.4	27	28.2	29.4	30.6	31.8	33	35	37	39	41	43	45
Elbow	24	24.5	25	26	27	28	29	30	31.5	33	34.5	36	37.5	39
Wrist	15.1	15.4	15.7	16.3	16.9	17.5	18.1	18.4	18.7	19	19.3	19.6	19.9	20.2
Thigh	48	49.5	51	54	57	60	63	66	69	72	75	78	81	84
Knee	36	36.5	37	37.5	39	40.5	42	43.5	45	46.5	48	49.5	51	52.5
Ankle	21.2	22.5	22.8	23.4	24	24.6	25.2	25.8	26.4	27	27.6	28.2	28.8	29.4

12.3 Measurement chart sizes 4 to 30 for women of average stature (height 165–170 cm/5′5″–5′7″)

12.8 Using a size chart to make clothing patterns

Please refer to Fig. 12.3 in this section.

12.8.1 Size types

I have given you a table that goes from a size 4 to a size 30. In general, Petite sizes tend to be at the smaller end of the size range, and Tall and Plus sizes at the larger end, but not necessarily so. A person may be tall and slim, or short and robust, so not strictly 'stock size'.

12.8.2 Collecting measurements

Use this chart as a helpful point of reference, and not an inflexible instruction. It is based on my own experience working in ladies' clothing in the UK in the recent past, and so is up to date, but also relies heavily on verification by the taking of measurements and observing body shapes and sizes all my working life.

By my side I have a whole jumble of notes and key measurements that I have collected along the way to which I have referred with gratitude. Every chance you get to measure someone, take it. While you have a willing someone, take those extra measurements that you have found hard to define. I have measured countless neck circumferences of women and have yet to find a correlation to graded size charts. Some have slim necks, some do not. It seems to depend more on body type and heredity than anything else. Use your own observation and some common sense rather than a blind acceptance of the size chart when you are making patterns. There will be other size charts that will differ with regard to the size nomination that I have used, for the reasons explained above.

12.8.3 Standards and variations

British Standards have tried to define and categorise garment sizes ever since the 1950s national size survey, but today such influences as changing ethnic mix, a general trend towards bigger people and plastic surgery make a one-size-spec-fits-all quite absurd. Regardless of national and international size specifications, for as many Pattern Cutters and Graders that there are, there will be as many opinions as to the correct measurements. This is a minefield that you have to navigate. Make your own observations and decisions.

Fortunately most fashion houses have a well-defined set of measurements that they adhere to, which is a great help as it defines the figure type and size range that you are working to. These garments will not fit the whole population. They are not intended to but are designed to be appropriate to the target customer.

12.9 References

Butterick *The World's First Name in Sewing Patterns. Butterick history*. Available from: http://butterick.mccall.com/butterick-history-pages-1007.php (accessed 29 September 2010).

Butterick Publishing Company *Making Smart Cloths: Modern Methods in Cutting, Fitting and Finishing*. New York, Butterick Publishing Company, 1930.

Kunick Philip, *Sizing, Pattern Construction and Grading for Women's and Children's Garments*. London, Philip Kunick Ltd, 1967.

Morris F. R., *Ladies' Garment Cutting and Making: A Standard Textbook Giving Instruction in all Branches of Ladies' Garment Cutting, Dress Cutting, Modelling, and Practical Tailoring*. London, The New Era Publishing Co. Ltd, 1934.

Richards Philip H., *Dress Creation*. London, Richard Arling Ltd, 1937.

Seligman Kevin L., *Cutting for All! The Sartorial Arts, Related Crafts, and the Commercial Paper Pattern: A Bibliographic Reference Guide for Designers, Technicians, and Historians*. Southern Illinois University Press, 1996.

SizeUK London 2002 http://www.size.org/ (accessed 9 September 2010).

Waugh Norah, *The Cut of Women's Clothes, 1600–1930*. London, Faber and Faber, 1968, re-issue 1994.

Women's Measurements and Sizes. London, Her Majesty's Stationery Office, 1957.

Chapter 13

How do I? The functions and menus that you need to know in Lectra Modaris indexed by the required action

Abstract: In this chapter the most commonly used functions are explained by task, allowing a function search based on what action is required. It can then be crossed-referenced with the function menus in Chapters 6 and 7.

Key words: How do I – Add, Create, Copy, Delete, Measure, Move, Undo, Zoom.

13.1 Introduction

This list is ordered by task, answering a question of 'How do I?' Rather than explaining what a function does, this chapter tells you what function you need for a specific task. For more detailed explanations about the functions, refer to Chapters 6 and 7 where they are set out in menu order.

13.2 Create or add

13.2.1 Create or add points

F1 – Create point First click on an existing point and then at the position of the new point required.

Shift + **F1** – **Create point** Use the same command but with **Shift** for a curved point and don't forget to turn on **Curved point** at the base of the screen to see them.

Relative point – known as drill holes in pattern cutting terminology, and placed within the pattern outline, so not on any line. **F1 – Relative point.** Click on any existing point first and then at the position required.

Drill hole – see **Relative point**

13.2.2 Create or add lines

F1 – Straight line Click at the start of the line, then click again to end the line.

F1 – Parallel line Click on line to be copied, then click on required position or enter a value in the measurement window.

F1 – Bezier for a curved line, hold **Shift** for curved points.

Shift + **F1 – Straight** will make a new line at a right angle to the existing line, useful for drawing a right angle when the base line isn't on the *x*- or *y*-axis.

Ctrl + **F1 – Straight** When creating a straight line it will keep it on the true horizontal or vertical – use the space bar to select which. It will also keep a perfect 45° angle. If you pull the line to any angle other than 90° it will lock onto the true bias line.

Grain line **F4 – Axis – Grain** Click at the position of the beginning of the line, hold **Shift** while clicking again at the end of the line. **Shift** will keep it on the true horizontal.

Comments line **F4 – Axis – Special axis** Click at the position of the beginning of the line, hold **Shift** while clicking again at the end of the line. Place wherever needed on a pattern piece.

13.2.3 Create sheets

New sheet: **Sheet – New sheet** (along the top of the screen). **N** on the keyboard.

New piece from an existing piece: **F4 – Seam** Click and drag in the section required. Selection will turn a luminous blue. Right click to finish the action and a new sheet will appear, maybe off screen so **J/j** or **8** to bring into view. The **Page down** shortcut is helpful since it shows only the piece to be extracted.

13.2.4 Create new model

New Model file from an existing model: Open the pattern to be copied. Use **File – Save as** and save to a new number. The first save will not show a prompt screen. If the screen appears showing **'this file already exists'** abort and chose a new unique number. Once saved, the numbering on each individual sheet needs to be amended. Use **Ctrl** + **A** to select all**.** All of the pattern pieces will become highlighted indicated by the squares at the corners of the pages. Go to **Edit – Rename** and a new small window will open. All of the pieces can now be renamed. It will ask for the existing name / number in the top box, then enter the new name in the lower box. Press Enter to complete the action. Save the file.

13.2.5 Add notches

F2 – Notch Click on a point to add a notch (do not add to curved points) or add anywhere along a line. If you add a notch along a line but not to a specific point, it will appear with a loop around the line, allowing movement along the line (**F3 – Reshape**) without changing the line in any way. To attach the notch to the line, use **F3 – Insert point**

Notch direction: **F2 – Orientation**, then click on the notch itself and drag in the direction required. Two notches can be attached to a point if required and set in different directions.

13.2.6 Add seam allowance

F4 – Line seam, then enter the seam allowance values in measurement box. Generally the seam allowance will be the same amount at both ends of the seam line, but the function allows for a different amount at each end.

13.2.7 Add text

Edit – Edit (along the top of the screen) to add text in text boxes

To add text on a piece **F4 – Axis – Special** to create a line, then select **Edit – Edit** and type text onto line making sure that the cursor remains purple.

13.2.8 Create or add a size range

F7 – Imp EVT to add an existing size range to a new pattern. See Chapter 11 for how to create a new size range.

13.2.9 Create a variant

F8 – Variant A new window will appear top left of screen, enter the variant name and press Return. The variant may be off screen, so press **J/j** or **8** to bring everything into view. If the variant already exists, use the same command but instead of entering a name in the name window, click directly onto the yellow variant sheet. More than one variant can be created and contained within one model.

13.3 Copy

13.3.1 Copy a line

There are two options: **Parallel** will only work within the same sheet, while **Duplicate** can also be copied to another sheet.

F1 – Parallel line Click on line to be copied, then click on required position or enter a value in the measurement window. If a curved line is copied the new line will distort to remain parallel to the original. To copy or duplicate the line retaining its original shape use **Duplicate**.

F1 – Duplicate Click on the line to be copied, then click again at required position. To duplicate a selection of lines hold **Shift** while making the selection with the right mouse button.

13.3.2 Copy a piece

Sheet – Copy (along the top of the screen) Click on required sheet, move cursor slightly. Left click and a new sheet will be created. It may not be visible, so **J/j** or **8** to bring into view.

13.3.3 Copy a pattern

See Section 3.2.4 and Chapter 8, Section 8.3.2 – **Rename**.

13.4 Delete

When you use the **Sheet – Delete** function a skull and crossbones icon will appear to warn that the function is active. To exit the function, go to **Selection** situated at the top of the right-hand column of the tools menus (this will in fact exit any selected function).

Point **F3 – Deletion**
Line **F3 – Deletion**
Sheet **Sheet – Delete**

13.5 Digitise

Pattern pieces **F1 – Digit**. See Chapter 4.

13.6 Display

Title Blocks: **Display – Title blocks** (along the top of the screen).

Curve points: turn on the **Curved points** tab at foot of screen.

Pattern outline: turn on **Cut piece** tab at foot of screen.

Toolbox icons: **Ctrl** + **S** will change from words to images.

Seam allowances: seam allowances can be defined inside or outside of the solid working line. When digitising a pattern that has the seam allowance already added there is no need to further define the seam allowance, but is by default to the inside of the piece. To define the amount use **F4 – Line seam**. The seam, defined by a dashed line, will only be visible when the **Cut piece** tab is activated. For a pattern digitised without the seam allowances, **F4 – Exchange** will swap the definition from within the solid line to the outside of the line.

13.7 Join

Two sheets: **F8 – Marry** Click first on the sheet you want to move, then onto the receiving sheet. This action allows one sheet to be within another temporarily and can be returned by **Divorce**. **Shift** + **F8 – Marry** will turn the piece over.

Two sheets permanently: **F8 – Assemble, Sheet – Copy** then **F4 – Seam.** See Chapter 7, Section 7.2.7 **F4 – Seam**.

Two lines: **F3 – Merge** changes the status of a point and therefore the existing adjacent lines. By merging the corner or end (Square) point to become a straight or type 1 point, two lines become one defined line. There are three types of points:

- square corner points;
- type 1 points (white) which are essentially non-curve points so keep the section between two adjacent type 1 points straight; note – when either of these points are graded they turn blue;
- red curve points.

F3 – Section will change the point status to a corner or end point, thus defining seam sections. This allows seam allowances to be different along different seam line sections.

F3 – Adjust 2 lines is used to shorten and clip lines together. Click on one line then the other, the first line will shorten to the second line, then click the second line followed by the first line and the two lines will be shortened to the same point.

13.8 Move

13.8.1 Move a point

F3 – Reshape Click on the point to be moved, and reposition.

F1 – Ali2pts To move a point so that it is exactly on the same axis as the first point. Click on the first point. An arrow will appear showing the direction the point will be moved. Click on the second point. To choose the opposite axis, use the space bar.

13.8.2 Move a line or piece

Line: **F3 – Move** (use **Pin** first, and select – right click – the line first).

Piece or sheet: **End** (key) – use with mouse to click on selected piece and move as required.

Married piece: **F8 – Move marriage**. See also **F8 – Pivot**: remember the piece will pivot from the last selected point.

13.9 Moving around the screen

See Chapter 2, Section 2.6 for detailed explanations.

J/j brings all work into view. With **Modaris V6** only **j** re-sizes all of the work to fit the screen, and is needed when working on a selected new window using **7**.

8 – brings all sheets into view.

Home – to bring a single sheet to fill the screen. Alternative – **Current sheet**.

End – then left click and drag to move individual sheets around on the screen.

Page up or **page down** to scroll between sheets.

13.10 Select or choose

Point: **Right click** and selection will be highlighted.

Line: **Right click** and selection will be highlighted.

Section: **Right click** or **Right click** + **Shift** to make multiple selections. Alternatively **click and drag** a box around section.

13.10.1 Between options

Space bar – allows a choice when more than one option is possible.

For example, when using F1 – Align 2 points – is it the *x*-axis or the *y*-axis? Choose with the Space bar.

Some functions always have a multiple choice, for instance when measuring a seam, is it this seam or the one adjacent? Press the space bar to go from one to the other. The Space bar is a choice bar.

13.11 Separate

Two or more married pieces **F8 – Divorce.**

13.12 Undo

Ctrl + **Z** undoes the last command. The MOST useful command, so don't worry about making a mistake, just remember **Ctrl** + **Z.** My constant companion and not just for mistakes. By combining its use with the **Print** function at the bottom of the screen, it can be used like a tracing function. Marry a piece and then turn on the **Print** function. **Ctrl** + **Z** and the sheet is divorced from the main sheet leaving a shadow outline.

Ctrl + **W** will redo the last command.

13.13 Zoom

See also Section 13.9 Moving around the screen.

13.13.1 Zoom in

Single sheet: **Home** – brings a selected sheet to full screen.

Selected small area: **return key** + **mouse** to click and drag to select a specific area to enlarge.

Resize sheet: when sheet is too big for the image, due to marrying pieces. **Caps lock** + **a** is the command you need to resize a page back to its smallest. Useful after marrying pieces which leaves the page very large after **divorcing** the pieces.

13.13.2 Zoom out

All work to be visible **8**.

All work visible in selected **7** window only, **J/j** – make sure it is a lower case for newer versions.

Appendix 1

Pattern numbering

Chapter 3, Section 3.1.2 explains the way in which computer-made patterns are organised using **title blocks**. Each pattern needs a unique number or reference name used with a disciplined method for logging those references. However, a list of numbers doesn't help when looking for a particular style without knowing the number. It is useful then to record your own pattern sketches and numbers, the file pathway and the destination folder for ease of future access.

In the example below the client is a large department store that has a file within the Lectra/files destination. In addition to this quick reference file, it is also useful to have a file with a full page working drawing or sketch, with the file reference number and dates of creation and amendments. The example is followed by document blanks for your own use.

Pattern numbering for . . . ***BOBS – Rugby club kit***

File name and pathway . . . ***C: Lectra/Files/ Bobs***

Model No. or root	Date	Front view	Back view
*BOB*001	*1/2/20*	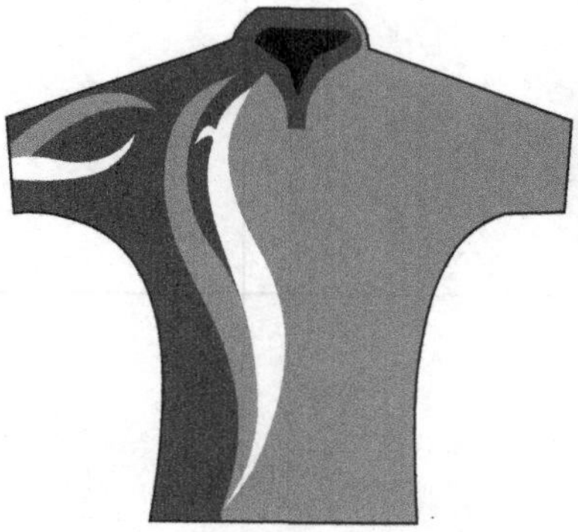	 AKUMA
*BOB*002	*5/2/20*		

Pattern numbering for ....

File name and pathway ....

Model no. or Root	Date	Front view	Back view
001			
002			
003			
004			
005			

Pattern numbering for ....

File name and pathway ....

Model no. or root	Date	Front view	Back view
006			
007			
008			
009			
010			

Pattern numbering for ....

File name and pathway ....

Model no. or root	Date	Front view	Back view
011			
012			
013			
014			
015			

Pattern numbering for ....

File name and pathway ....

Model no. or root	Date	Front view	Back view
016			
017			
018			
019			
020			

Appendix 2

Recommended abbreviations for naming pattern pieces in Lectra Modaris

Chapter 3, Section 3.5 explains the purpose of pattern piece naming and the role of the variant. When filling in the variant, pattern pieces need to be identified with a unique number or code. This will not affect the pattern as the system will, by default when creating a new sheet, automatically fill the code with the next available number. However, when searching for patterns, or more specifically for a particular piece, it helps to have a consistent reference name for your pattern pieces. The following are recommended. The list is not exhaustive and there are many occasions when there is no regular piece name. In this case create your own but be mindful of the codes already in existence.

Main cloth:

FR	Front
MF	Mid front
SF	Side front
CF	Centre front
RFR	Right front
LFR	Left front
FF	Front facing
RFF	Right front facing
LFF	Left front facing
BK	Back
MB	Mid back
SB	Side back
CB	Centre back
LBK	Left back
RBK	Right back
BNF	Back neck facing
TS	Top sleeve
US	Under sleeve
S	Sleeve
C	Cuff
TC	Top collar
UC	Under collar
ST	Collar stand
PB	Pocket bag
PO	Pocket
PF	Pocket facing
J	Jet
W	Welt
LWB	Left waistband
LWF	Left waist facing
RWB	Right waistband
RWF	Right waist facing
BWB	Back waistband
BWF	Back waist facing
WB	Waistband
ZG	Zip guard
F	Fly
B	Belt
L	Belt loops
TI	Tie
NB	Neck binding
AB	Armhole binding
STB	Strap binding
B	Binding

For all the above, if fused add an F after abbreviation i.e. FRF = Front fuse

Lining – always end in L:

FL	Front lining
RFL	Right front lining
LFL	Left front lining
SFL	Side front lining
BL	Back lining
RBL	Right back lining
LBL	Left back lining
SBL	Side back lining
TSL	Top sleeve lining
USL	Under sleeve lining
PL	Pocket lining
PBL	Pocket bag lining

Non-fusible interlining:

BAF	Back armhole nonfuse
FAF	Front armhole nonfuse

Fusible interlining – always end in F:

SBHF	Side back hem fuse
BHF	Back hem fuse
TSF	Top sleeve hem fuse
USF	Under sleeve hem fuse
SHF	Sleeve head fuse
SAF	Side under arm fuse

Markers – always end in M:

BM	Button marker
BHM	Buttonhole marker
FM	Fly marker
HM	Hook marker
PM	Pocket marker

Pattern fabric codes for use in the variant and Diamino:

Self	0
Fuse	1
Lining	2
Card	3
Contrast	4

Appendix 3

Keyboard commands by alphabet

It is very easy to activate a shortcut unintentionally, the most visible being the display functions such as G, display grid and h, the pin function. Once this is realised, the screen can be returned to your preference by de-selecting the function. While aiming for J, I have frequently hit k instead, and activated the associated parameters dialogue box, or N which will give you a new sheet. Again, knowing the reason for the sudden appearance of an unexpected object helps you to understand Modaris better and then easily return the working environment to your requirements. Any omissions in this list are because the shortcut is assigned to another level of Modaris. Some letters have no assigned function listed.

a	Sheet – Adjust
A	F1 – Align 2 points
Alt + a	F1 – Align 3 points
Ctrl + A	Selection – Select all
b	F1 – Create a bezier line
B	Set as 3D
Alt + b	Layers – View all layers
c	F2 – Notch
C	Selection – Notches filter
Alt + c	Display – Print
Ctrl + C	Sheet – Copy
d	F8 – Divorce
D	F3 – Move
Alt + d	F1 – Duplicate
Ctrl + D	F1 – Recover digit
e	F3 – Remove pin
E	Edit – Edit
Alt + e	F3 – Pin
Ctrl + E	File – Save
f	
F	F3 – Pins ends
Alt + f	Sheet – Arrangement record
g	Layers – Erase
G	Display – Grid
Alt + g	F6 – Launch easy grading
Ctrl + G	F6 – Control
h	F3 – Pin graded points
H	Display – Handles
Alt + h	F2 – Perpendicular
Ctrl + H	Display – Scale
i	Sheet – Sheet select
I	F1 – Intersection
j	Display – Arrange all
J	Display – Arrange all
Alt + j	F8 – Hide measures
Ctrl + J	File – Access paths
k	Parameters – Associated parameters
K	Create piece article
Alt + k	Sheet – Transparent
Ctrl + K	Display – Ftype
l	F8 – Length
L	F8 – Length measure
Ctrl + L	F8 – Seam length
m	F8 – Marry
M	F2 – Marking
Ctrl + m	F8 – Assemble
n	
N	Sheet – New
Ctrl + N	File – New
o	F4 – Seam
O	F4 – Piece seam
Alt + o	Display – Hide sym obj
Ctrl + O	File – Open

p	
P	Display – Curve points
Alt + p	F3 – Stretch
Ctrl + P	Display – Flat pattern
q	Quality area
Q	Selection – Filter
Alt + q	Selection – Flat pattern
r	F3 – Reshape
R	F6 – Equate
Alt + r	F1 – Relative
Ctrl + R	F1 – Digit
s	Selection
S	F1 – Semi-circular
Alt + s	F3 – Simplify
Ctrl + S	Configure – Icons / text
t	
T	F2 – Rectangle
Alt + t	F8 – Spreadsheet
Ctrl + T	File – Save as
u	F2 – Notch parameters
U	F4 – Line seam
Alt – u	F2 – Notch orientation
Ctrl + U	Display – Title blocks
v	F1 – Developed
V	File – Model validation
Ctrl + V	File – Export garment
w	Parameters – Actu. associated parameters
W	Parameters – Associated param close
Ctrl + W	Edit – Redo last command
x	F1 – Sym axis
X	F1 – Parallel
Alt + x	Layers – Display layers colours
Ctrl + X	File – Quit
y	F4 – Cut
Y	F1 – Symmetrize
Alt + y	F2 – Bisecting line
Ctrl + Y	File – Import BI garment
z	Sheet – Delete
Z	Selection – De-select
Alt + z	Selection – Sequence
Ctrl + Z	Edit – Undo last command

0	F1 – Straight line
1	Layers – Construction
2	Layers – Base
3	F4 – Import piece
4	F4 – Export piece
5	
6	Display – Show protected objects
7	Sheet – Selective visual
8	Sheet – Visual all
9	
Alt + 0	Layers – Hidden objects
Alt + 1	
Alt + 2	
Alt + 3	Layers – Display seam/cut lines
Alt + 4	F1 – Add point
Alt + 5	F3 – Attach
Alt + 6	F3 – Detatch
Alt + 7	F3 – Insert point
F1	Points and lines menu
F2	Notches, orientation and tools menu
F3	Modification menu
F4	Industrialsation and piece menu
F5	Derived pieces, folds and CAM menu
F6	Grading menu
F7	Evolution menu
F8	Measuring and assembly menu
F9	Display-cut piece same as F6-nest
F10	Display – sizes
F11	Selection – break sizes
F12	Selection – all sizes
“	F4 – axis – grain
$	F3 – lengthen
*****	F8 – variant
(	F3 – attach
)	F3 – detach
–	F3 – lengthen a straight line
+	F3 – merge
{	F2 – 2 cirlces tangent
]	F8 – move marriages
[	F8 – pivot
#	F8 – charts manager
Delete	Deletes lines and points
Home	Brings a single selected sheet to full screen

End	Used with the mouse to pick up and move a sheet
Page up/ down	Scroll between successive sheets
Number keypad	Enter specific number values
Movement keys	Move between value fields in a dialogue box
Enter	Use with mouse to 'Click and drag' a rectangle for enlargement
Space bar	The choice bar, when more than one option is available within a function
Shift	Used to vary a function
Control	Used to vary a function
Tab	Used for inserting existing files into the current file
Back space	Display – scale

Appendix 4

Lectra Modaris terminology glossary

Lectra terminology varies from traditional pattern cutting terminology, and what is easily understandable in one case may not be so obvious another. Computer programs often need additional terms for something that did not exist for card patterns, and sometimes different words are used in different contexts but actually mean the same thing. This glossary is intended to put some Lectra terminology into plain pattern cutting terms.

acd – analytical code (text box field). Taken as a whole, the Lectra CAD system has a variety of separate parts working with different file types which need to work together. The **acd** enables these separate parts to identify the various pattern pieces. For example when a piece is added to the **variant**, which is a **text file**, it needs to be identified with the **basic image** which is a **graphics file**.

BI – basic image, see pattern piece.

CAD – computer aided design.

CAM – computer aided manufacture.

Col: – collection (text box field).

Comment – a text box field allowing you to type in relevant text which can then be plotted out on your pattern. The comment box has a limited number of characters, but enough to make a short sentence.

Dialogue box – an interactive window in which to enter measurement values, names or numbers for the current function.

Diamino – the Lectra software for creating lay plans using patterns created on Modaris.

File name – see Pattern.

Gcd – analytical code (text box field).

Model, **Model file, Model name** – see Pattern.

Modaris – file extension **.mdl** or **.MDL** – the software for pattern making.

Pattern – the collection of individual pattern pieces that together make up a **Pattern**, is also known as, or referred to by Lectra in various ways as, the:

- **Model**,
- **Model file**,
- **Model name**,
- **File name**,
- **Root**.

On the **model identification sheet** title block it is the – **name**.

On the **variant sheet** title block it is the – **root**.

On the **individual piece sheet** title block it is also – **root**.

By default the **variant** and **piece** sheets take their information from the **model identification sheet**, hence – **root**.

Pattern piece – the single shape or template of one pattern piece is also known as or referred to by Lectra as:

- **BI**, which stands for **basic image**,
- **Basic image** is contained on a **Sheet**,
- **Piece name**,
- **Piece article** – when adding to a variant.

When the pattern is sent to **Diamino** for marker making, the software needs to be able to identify, and import the **basic image (BI)**.

Model identification sheet – the first sheet to be created automatically when a new model is created.

Name – see Pattern.

Piece article, Piece name – see Pattern piece.

Root – see Pattern.

Title blocks – the yellow text fields sections on the left and along the bottom of the pieces sheets of a model.

Variant – the spreadsheet containing the pattern information and instructions for cutting. On a traditional pattern this information would have been written on each pattern piece.

Index